ESD IN SILICON INTEGRATED CIRCUITS

DESIGN AND MEASUREMENT IN ELECTRONIC ENGINEERING

Series Editors

D. V. Morgan
School of Electrical, Electronic and Systems Engineering
University of Wales College of Cardiff,
Cardiff, UK

H. L. Grubin
Scientific Research Associates Inc.
Glastonbury, Connecticut, USA

THYRISTOR DESIGN AND REALIZATION
P. D. Taylor

ELECTRONICS OF MEASURING SYSTEMS
Tran Tien Lang

DESIGN AND REALIZATION OF BIPOLAR TRANSISTORS
Peter Ashburn

ELECTRICAL CHARACTERIZATION OF GaAs
MATERIALS AND DEVICES
David C. Look

COMPUTERISED INSTRUMENTATION
Tran Tien Lang

PRINCIPLES AND TECHNOLOGY OF MODFETs
VOLUMES 1 AND 2
H. Morkoç, H. Unlu, G. Ji

RELIABILITY OF GALLIUM ARSENIDE MMICs
Edited by Aris Christou

INTEGRATING QUALITY AND RELIABILITY INTO MICROELECTRONICS MANUFACTURING
Aris Christou

DESIGN OF SMALL ELECTRICAL MACHINES
E. S. Hamdi

IMAGE BASED MEASUREMENT SYSTEMS
F. van der Heijden

ESD IN SILICON INTEGRATED CIRCUITS
Ajith Amerasekera
Charvaka Duvvury

ESD IN SILICON INTEGRATED CIRCUITS

Ajith Amerasekera
Charvaka Duvvury
Texas Instruments Inc.
Dallas, Texas, USA

JOHN WILEY & SONS
Chichester • New York • Brisbane • Toronto • Singapore

Copyright © 1995 by John Wiley & Sons Ltd,
Baffins Lane, Chichester,
West Sussex PO19 1UD, England

National   01243 779777
International (+44) 1243 779777

Reprinted May 1996

Other Wiley Editorial Offices

John Wiley & Sons, Inc., 605 Third Avenue,
New York, NY 10158-0012, USA

Jacaranda Wiley Ltd, 33 Park Road, Milton,
Queensland 4064, Australia

John Wiley & Sons (Canada) Ltd, 22 Worcester Road,
Rexdale, Ontario M9W 1L1, Canada

John Wiley & Sons (SEA) Pte Ltd, 37 Jalan Pemimpin #05-04,
Block B, Union Industrial Building, Singapore 2057

British Library Cataloguing in Publication Data

A catalogue record for this book is available from the British Library

ISBN 0 471 95481 0

Typeset in 10/12pt Times by Techset Ltd, Salisbury, Wiltshire
Printed and bound in Great Britain by Bookcraft (Bath) Ltd
This book is printed on acid-free paper responsibly manufactured from sustainable forestation,
for which at least two trees are planted for each one used for paper production.

This book is dedicated to:
our parents
Rohan and Aloma Amerasekera
Rao and Sarada Duvvury
and our wives
Anoma Amerasekera
Vasu Duvvury

CONTENTS

SERIES PREFACE

The crucial role of design in the engineering industry has been increasingly recognized over recent years, with particular emphasis being placed on this aspect of engineering in first and higher degree work as well as continuing education.

This series of books concentrates on fundamental aspects of design and measurement in electronic engineering and involves an international authorship. The authors are sought from scientists and engineers who have made a significant contribution in their field. The books in the series cover a range of topics at research level and are primarily intended for research and development engineers wishing to gain detailed specialist knowledge of design and measurement in a particular area of electronic engineering. It is asssumed that, as a starting point, the reader will have a background degree or equivalent qualification in electrical and electronic engineering, physics or mathematics. In the series no attempt is made to provide preliminary background material but rather the texts move directly into the design aspects.

Professor D. V. Morgan
Dr H. R. Grubin

PREFACE

Electrostatic Discharge (ESD) phenomena in integrated circuits (IC) have grown in importance as technologies shrink to below 0.5 μm dimensions and the number of transistors on a single chip approach the 5 million mark. The phenomena related to ESD events in semiconductor devices take place outside the realm of normal device operation. Hence, the physics governing this behavior are not typically found in general textbooks on semiconductors. Similarly the circuit design issues involve non-standard approaches which are not covered in general books on electronic design. Over the years there has been much work done in the areas of ESD circuit design and the physics involved, most of which has been published in a number of papers and conference proceedings. We feel that the field has reached sufficient maturity that the work can be combined and published as a book. This book covers the state-of-the-art in circuit design for ESD prevention as well as the device physics, test methods and characterization. We also include case studies showing examples of approaches to solving ESD design problems.

The book is intended for engineers and scientists working in the field of IC circuit design and transistor device design. In addition the basics presented in this book should also appeal to graduate students in the field of semiconductor reliability and device/circuit modeling. As the problems associated with ESD become significant in the IC industry the demand for graduates with a basic knowledge of ESD phenomena also increases. We hope that this book will help students to meet the demands of the IC industry in terms of understanding and approaching ESD problems in semiconductor devices.

As anyone involved in writing a book or review article is aware, the contents of such documents summarize the excellent work of many researchers in this field. There are many companies and research institutes that have made it possible to understand and solve the majority of ESD problems in ICs. We do not have room to mention them all but some of the companies that have been particularly active in recent years are Texas Instruments, Philips Semiconductors (Holland), AT&T Bell Labs, IBM, RCA David Sarnoff Labs and Intel. Research Institutes that have made significant contributions in recent years are Sandia National Labs, Clarkson University, Clemson University, the University of California in Berkeley, the University of Western Ontario in Canada, the University of Illinois at Urbana-Champaign, Loughborough University in England, Twente University in The Netherlands, the Technical University of Munich in Germany and IMEC in Belgium.

We acknowledge the foresight of Lionel White in initiating active research into ESD at Texas Instruments in the early 1980's. The results of the programs initiated by him have benefitted our understanding of the ESD phenomena and design techniques with consequences which have extended beyond Texas Instruments. On a personal level, we would particularly like to thank Robert Rountree, Thomas Polgreen and Amitava Chatterjee for their many contributions to our knowledge and understanding, both at the circuit design and at the device level. Ping Yang and William Hunter have provided excellent technical guidance during the evolution of the work on ESD, and without their management support this work would not have been undertaken. Many of our colleagues here at Texas Instruments have done the ground work which has helped us to expand our understanding in this area. We are especially grateful for the contributions of Kuen-Long Chen and David Scott in this respect. In the area of device physics and modeling, the contributions of Mi-Chang Chang, Kartikeya Mayaram, Jue-Hsien Chern and Jerold Seitchik have been invaluable. We have also been fortunate to interact closely with researchers from other organizations in the period that we have been involved in this work. In particular, we would like to thank Jan Verweij, Fred Kuper and Leo van Roozendaal from Philips Semiconductors in Holland; Andy Franklin, Vincent Dwyer and David Campbell from Loughborough University, England; and Henry Domingos at Clarkson University, for their contributions over the years. Finally, we thank Anoma for going through the painstaking task of proof-reading the entire manuscript and ensuring that it is readable and that the grammar is defect-free.

Ajith Amerasekera
Charvaka Duvvury
Dallas, November 1994

1

INTRODUCTION

1.1 BACKGROUND

The phenomenon of *Electrostatic Discharge* or ESD conjures up images of lightning strikes or the sparks that leap from one's fingertips when touching a door knob in dry weather. The sparks are the result of the ionization of the air-gap between the charged human body and the zero potential surface of the door knob. Clearly a high voltage discharge takes place under these circumstances with highly visible (and sometimes tangible) effects. In the semiconductor industry, the potentially destructive nature of ESD in Integrated Circuits (IC) became more apparent as semiconductor devices became smaller and more complex. The high voltages result in large electric fields and high current densities in the small devices which can lead to breakdown of insulators and thermal damage in the IC. The losses in the IC industry due to ESD can be substantial if no efforts are made to understand and solve the problem [Wagner93]. Figure 1.1 shows the distribution of failure modes observed in silicon ICs, and ESD is observed to account for close to 10% of all failures [Green88]. The largest category is that of Electrical Overstress (EOS) of which ESD is a subset. In many cases failures classified as EOS could actually be due to ESD which would make this percentage even higher [Merrill93].

The significance of ESD as an IC failure mode has led to concerted efforts by IC manufacturers and university research workers in the US, Europe and Japan to study the phenomena. Progress has been made in understanding the different types of ESD events affecting ICs which has enabled test methods to be developed to characterize their ESD robustness as has been documented in earlier books by Bhar and McMahon [Bhar83] and Greason [Greason87]. ESD prevention programs have been put in place during IC manufacturing, testing and handling which have reduced the build-up of static and the exposure of ICs to ESD. Studies have been made of the nature of destruction in IC chips and, based upon this work, techniques for designing protection circuits have been implemented which has made it possible for the present generations of complex ICs to be reasonably ESD robust.

The introduction of each new generation of silicon technology results in new challenges in terms of ESD capability and protection circuit design. Figure 1.2 shows

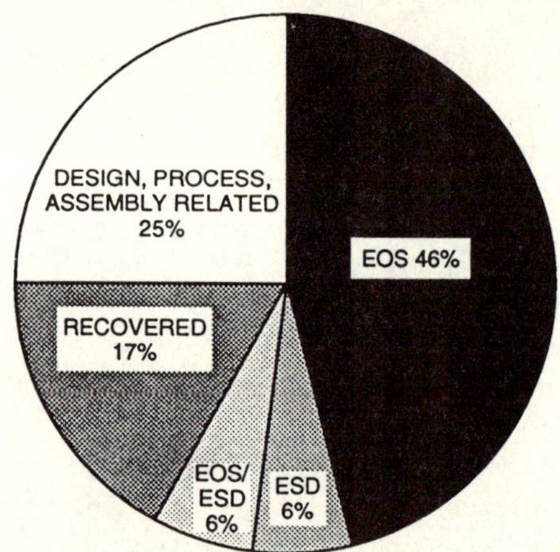

Figure 1.1   Distribution of failure modes in silicon ICs. ESD accounts for approximately 10% with EOS responsible for close to 50% of the failures (after [Green88]).

how ESD performance for specific protection circuits has changed over periods of time. Initially the ESD performance improves as the circuit designs mature and problems are solved or debugged. After a certain time the technology changes (i.e. LDD, salicides) cause the circuit to no longer function to its original capability and new protection techniques are needed to restore good ESD performance. As a result, the ESD issue is expected to remain important for the foreseeable future of IC technology. This is equally applicable to ICs manufactured in silicon or GaAs, whether they are CMOS, bipolar, MESFETs or heterojunction devices. The increased usage of CMOS ICs in automotive environments requires very high ESD

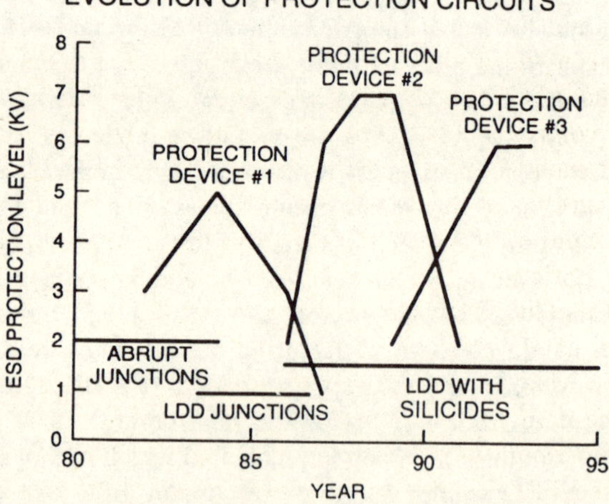

Figure 1.2   ESD protection levels as a function of time. As technologies change, new protection devices are needed to reach the same ESD levels as before.

protection levels which places an even higher demand on the design of protection circuits.

The speed with which new technologies are introduced has reduced the available time for protection circuit development. In fact it is becoming more and more important to design circuits which can be transferred into the newer technologies with minimum changes. Hence, it is necessary to understand the main issues involved in ESD protection circuit design and the physical mechanisms taking place in order to ensure that the design can be scaled or transferred with minimum impact on the ESD performance.

It is the purpose of this book to provide an introduction to the basic mechanisms involved in ESD events, the physical processes taking place in the semiconductor, and the design and layout requirements for best ESD performance. We have focused on silicon since most volume IC production uses this material. The importance of building-in reliability demands design approaches that include ESD robustness as part of the technology roadmap. The design and optimization of ultra-small transistors (sub-0.25 μm) and the associated processes increasingly use simulation tools prior to committing to silicon. This trend makes modeling and simulation of ESD effects in the protection device important issues, especially in the coming generations of ICs, and we discuss the main approaches here. The book is aimed at providing an overall picture of the issues involved in ESD protection circuit design and analysis. It is intended to provide a basis in this field for circuit design and reliability engineers as well as process and device design engineers who have to deal with ESD in integrated circuits.

1.2 THE ESD PROBLEM

ESD is the transient discharge of static charge whether it arises due to human handling or contact with machines. The mathematics of the generation of static electricity have been presented in some detail in previous works [Bhar83] [Greason87] and the interested reader is referred to those works for more on the fundamentals of electrostatics. In a typical work environment a charge of about 0.6 μC can be induced on a body capacitance of 150 pF, leading to electrostatic potentials of 4000 volts or greater. Any contact by the charged human body with a grounded object such as an IC pin can result in a discharge for about 100 nanoseconds with peak currents in the ampere range. The energy associated with this discharge could mean failure to electronic devices and components. Typically, the damage is thermally initiated in the form of device or interconnect burn-out. The high currents could also lead to on-chip voltages which are high enough to cause oxide breakdown in thin gate MOS processes.The latter form of damage requires a large amount of energy. Many semiconductor devices can be damaged even at a few hundred volts but the damage is too weak to be detected easily, resulting in what is known as *walking wounded* or *latency effects* [McAteer82]. A device can be exposed to undetected ESD events starting in the fabrication area during process [Hill85] and extending through the various manufacturing stages up to the system level. Thus, precautions to suppress ESD become important through all phases of an IC's life.

As mentioned earlier, ESD is actually a subset of the broad spectrum of Electrical Overstress, or EOS, where the EOS family includes lightning and Electromagnetic Pulses (EMP). EOS, in general, commonly refers to events other than ESD that encompass time scales in the microsecond and millisecond ranges compared to the 100 ns range associated with ESD. EOS events can occur due to electrical transients at the board level or the system level. They can also occur during device product engineering characterization or during the burn-in test. Although much of the reliability focus has been on ESD, EOS is increasingly considered to be a major issue demanding more attention as it becomes a significant failure mode in the IC industry. This book will deal only with the ESD-related phenomena although much of the device physics and analytical modeling discussed here will be equally applicable to EOS stress conditions.

1.3 PROTECTING AGAINST ESD

The main ESD problem in a wafer fabrication area is static charge generation which should be suppressed. The prevention methods could be the use of antistatic coatings to the materials or the use of air ionizers to neutralize charges. The damage due to human handling can be reduced by the proper use of wrist straps for grounding the accumulated charges and shielded bags for carrying the individual wafers. Static control and awareness are two important programs to combat ESD in the semi-conductor manufacturing environment [McAteer79] [Dangelmayer85].

As a second step to reduce ESD effects, protection circuits are implemented within the IC chip [Lindholm85]. With effective protection circuits in place, the packaged device can be handled safely from device characterization to device application. However, the packaging procedure itself can cause serious damage and thus antistatic precautions are also needed during the wire bonding and assembly phases. Even with good protection circuits, devices are not necessarily immune to ESD once they are on the circuit boards. Other forms of ESD from the charged boards are possible. Thus ESD precautions are important during system assembly as well.

Finally, the implementation of effective on-chip protection is a continuous learning experience. Even if not very effective, a relatively weak protection circuit is better than none. A good protection design would be capable of surviving the ESD event and protect the internal transistors connected to the IC pin. It is a challenging task to design effective protection circuits and several design iterations can be required to optimize them. This is mainly because no reliable modeling programs are available to date and the designer thus has to rely on empirical studies on test structures in silicon. However, there has been considerable progress in recent years in device modeling for ESD. The eventual goal is that ESD protection circuits can be simulated and even optimized for a predictable level of ESD protection performance using standard circuit simulation tools such as SPICE.

1.4 OUTLINE OF THE BOOK

ESD phenomena in ICs are outside the usual operating ranges of the devices being stressed. The behavior of semiconductor circuit elements during an ESD event is, therefore, not covered by standard texts on semiconductor device physics. However, a general understanding of this behavior can be obtained from numerous publications on the high current behavior of bipolar devices [Ghandhi77]. Similarly, circuit design and layout for ESD robustness require particular guidelines which have evolved though years of experimental work in this field. The same is true for test methods and characterization. In this book we have attempted to present coverage of all these aspects which would enable the reader to gain a broad understanding of ESD in ICs and the main issues involved in improving ESD performance. The book draws from a large publication base in this area, the bulk of which is available through the *Proceedings of the EOS/ESD Symposium* which is held annually and deals with all areas of ESD. Much of the detailed understanding of ESD in ICs has been presented at this Symposium. Brief outlines have been presented in review papers which have presented the state-of-the-art regarding ESD at the time of publication[Amerasekera92][Duvvury93]. The book consists of eight chapters and an outline of the contents of each chapter is given below.

Chapter 2 presents the details of the ESD phenomena and the appropriate test methods. These phenomena are described in terms of the voltages, currents and pulse durations while the test methods are described in terms of the simulations of the events arising from the different stress models. The test methods, shown to approximate the phenomena, consist of the Human Body Model to represent the human handling, the Machine Model to emulate machine contact, and the Charged Device Model to determine the effects of field-induced charging of the packaged IC. The issues dealing with the accuracy of these models and the commercial testers available to simulate them are also discussed.

To understand the mechanisms of device failures and operation of the semiconductor protection devices under the high current, short duration ESD pulses, the device physics behind these will be considered in Chapter 3. The protection device design requires an understanding of the physics involved in resistors, reverse-biased *pn* diodes, the parasitic *npn* operation of an nMOS transistor, or the latchup operation of a *pnpn* device.

In Chapter 4, the design requirements for effective protection circuits that can perform without degrading the IC chip functions will be discussed. For example, a protection at the input should not affect the gate oxide reliability, an output protection should have no impact on the output buffer performance, and neither of these should result in an increase in the leakage current in the chip. The approach taken here will be to demonstrate a synthesis of the protection circuit design needs while considering the optimum design compatible with complex internal IC chip current paths during ESD, or the function of the chip – that is, whether it is floating substrate DRAM or a grounded substrate logic chip. Each individual protection element will be discussed separately first before combining them to form effective protection schemes. Just as important as the protection design is its implementation. The layout of a protection device plays a crucial role in its effectiveness. Both the design and layout techniques are discussed in Chapter 4. As will be demonstrated, effective protection circuit schemes can perform far below the expected level, mainly due to poor implementation. The chapter will focus on the design practices and guidelines for the protection design layouts.

Even after an effective design and layout, the full ESD robustness of the IC cannot always be guaranteed. To illustrate these points, the main failure modes observed in advanced silicon ICs will be discussed in Chapter 5, together with case studies related to the effects of design and layout on ESD performance. This analysis involves a thorough stress methodology for characterization and a full study of the failure modes. Several actual case studies will be presented that indicate the common, and sometimes more bizarre, ESD problems. A brief summary of the failure analysis techniques useful for ESD as well as the post-stress failure criterion will be reviewed.

In Chapter 6, a review of the modeling techniques based on the high current behavior of the protection circuits will be given. These will look at the approaches used in analytical and numerical modeling of the ESD phenomena in semiconductor devices. Emphasis is on the ability of these methods to predict device behavior under design and process variations. This is essentially still a new field and a lot of work is currently being done to uncover the underlying mechanisms involved and identify the main predictive indicators to be used [Amerasekera94]. The eventual goal of these programs is the capability to develop and evaluate high performance ESD protection circuits in new processes using simulation techniques.

The development of newer protection techniques is needed because of the degradation for the existing protection devices with advances in process technologies as shown in Figure 1.2. In many cases, process dependence of ESD performance can frustrate any attempts to achieve the specified ESD levels for the product. Chapter 7 discusses some of the principal aspects related to process effects, such as the impact of LDD junctions or salicided diffusions on ESD performance. The specifics of the process effects and methods to monitor these process effects will be reviewed. Besides the process, the effects of the package type and the package size will also be briefly discussed in this chapter.

Finally, in Chapter 8 a summary of the main issues will be given together with an evaluation of the present state-of-the-art with regard to protection techniques and future ESD requirements and directions for further work in this area.

REFERENCES

[Amerasekera92]   A. Amerasekera, J. Verwey, 'ESD in Integrated Circuits', *Qual. and Rel. Eng. Int.*, 8, p. 259–272, 1992.

[Amerasekera94]   A. Amerasekera, J. Seitchik, 'Electrothermal Behavior of Deep Submicron nMOS transistors under high current snapback (ESD/EOS) conditions' in *Tech. Dig. IEDM*, p. 446–449, 1994.

[Bhar83]   T. N. Bhar, E. J. McMahon, *Electrostatic Discharge Control*, New Jersey: Hayden, 1983.

[Dangelmayer85]   G. T. Dangelmayer, E. S. Jesby, 'Employee Training for Successful ESD Control,' in *Proc. 7th EOS/ESD Symposium*, p. 20–23, 1985.

[Duvvury93]   C. Duvvury, A. Amerasekera, 'ESD: A Pervasive Reliability Concern for IC Technologies', *Proceedings IEEE*, 81, p. 690–702, 1993.

[Ghandhi77]   S. K. Ghandhi, *Semiconductor Power Devices*, New York: Wiley, 1977.

[Greason87]   W. D. Greason, *Electrostatic Damage in Electronics: Devices and Systems*, London: Research Studies Press, 1987.

[Green88]   T. Green, 'A Review of EOS/ESD Field Failures in Military Equipment', in *Proc. 10th EOS/ESD Symposium*, p. 7–14, 1988.

[Hill85]   H. Hill, D. P. Renaud, 'ESD in Semiconductor Wafer Processing', in *Proc. 7th EOS/ESD Symposium*, p. 6–9, 1985.

[Lindholm85]   A. W. Lindholm, 'A Case History of an ESD Problem', in *Proc. 7th EOS/ESD Symposium*, p. 10–14, 1985.

[McAteer79]   O. J. McAteer, 'An Effective ESD Awareness Training Program', in *Proc. 1st EOS/ESD Symposium*, p. 1–3, 1979.

[McAteer82]   O. J. McAteer, R. E. Twist, R. C. Walker, 'Latent ESD Failures', in *Proc. 4th EOS/ESD Symposium*, p. 41–48, 1982.

[Merrill93]   R. Merrill, E. Issaq, 'ESD Design Methodology', in *Proc. 15th EOS/ESD Symposium*, p. 233–237, 1993.

[Wagner93]   R. G. Wagner, J. Soden, C. F. Hawkins, 'Extent and Cost of EOS/ESD Damage in an IC Manufacturing Process', in *Proc. 15th EOS/ESD Symposium*, p. 49–55, 1993.

2

ESD PHENOMENA AND TEST METHODS

2.1 INTRODUCTION

Voltages ranging from 100 V to 20 kV can be generated as a result of triboelectric charging of materials. Table 2.1 [Moss82] describes the triboelectric series showing the nature and extent of charge that can be carried by a variety of materials. It can be seen that air is capable of carrying the most positive charge, followed by human skin, rabbit fur and glass. Silicone rubber, teflon and silicon carry the most negative charge. Table 2.2 gives an indication of the amount of static voltage that can be generated by a number of common functions from walking to aerosol spraying. The magnitude of the electrostatic voltage is a function of the relative humidity, and higher voltages are generated with lower humidity.

In integrated circuit manufacturing and handling environments, there are three principal sources of electrostatic charging and discharging (ESD). The first and most common to date is that due to human handling. A person crossing a synthetic floor surface can accumulate up to 20 kV according to Table 2.2. This voltage is discharged when the person touches an object which is sufficiently large that most of the charge exchange will be from the person to the object. Such an object is effectively at ground. Many people have commonly experienced the phenomenon when leaving their cars, and the spark which extends between their hands and the car body or door handle is an example of the discharge taking place. The amount of energy involved is relatively small, of the order of microjoules, and causes only minor discomfort. However, if that discharge takes place through an integrated circuit, the currents and the energy dissipated are large enough to cause damage due to the very small dimensions of the semiconductor structures. The discharge occurs in a very short duration, typically of the order of 10 ns to 100 ns, with currents ranging from 1 A to 10 A depending on the conditions of the discharge. The second source of ESD is that which takes place in automated test and handling systems. The equipment can accumulate static charge due to improper grounding, which is then transmitted through the IC when it is picked up for placement in the test socket or

Table 2.1   Electrostatic triboelectric series [Moss82].

ELECTROSTATIC TRIBOELECTRIC SERIES

MOST POSITIVE (+) ⟶	AIR
	HUMAN SKIN
	ASBESTOS
	FUR (RABBIT)
	GLASS
	MICA
	HUMAN HAIR
	NYLON
	WOOL
	SILK
	ALUMINUM
	PAPER
	COTTON
	STEEL
	WOOD
	SEALING WAX
	HARD RUBBER
	NICKEL, COPPER
	BRASS, SILVER
	GOLD, PLATINUM
	ACETATE FIBRE (RAYON)
	POLYESTER (MYLAR)
	CELLULOID
	ORLON
	POLYSTYRENE (STYROFOAM)
	POLYURETHANE (FOAM)
	SARAN
	POLYETHYLENE
	POLYPROPYLENE
	POLYVINYL CHLORIDE (VINYL)
	SILICON
	TEFLON
MOST NEGATIVE (-) ⟶	SILICONE RUBBER

Table 2.2   Static voltages [Moss82].

STATIC VOLTAGES AS A FUNCTION OF HUMIDITY	20% RH (kV)	80% RH (kV)
WALKING ACROSS VINYL FLOOR	12	0.25
WALKING ACROSS SYNTHETIC CARPET	35	1.5
ARISING FROM FOAM CUSHION	18	1.5
PICKING UP POLYETHYLENE BAG	20	0.6
SLIDING STYRENE BOX ON CARPET	18	1.5
REMOVING MYLAR TAPE FROM PC BOARD	12	1.5
SHRINKABLE FILM ON PC BOARD	16	3
TRIGGERING VACUUM SOLDER REMOVER	8	1
AEROSOL CIRCUIT FREEZE SPRAY	15	5

carrier. The accumulated charge may be higher than in the cases for human handling, but the static voltages are typically lower. The discharge is a short duration current pulse of very high magnitude. The third possibility is that the IC itself is charged during transport or because of contact with a highly charged surface or material. The IC remains charged until it comes into contact with a grounded surface such as a large metal plate or a test socket. It is then discharged through its pins and the large currents in the internal interconnect can result in high voltages inside the device. These voltages can cause damage to the very thin dielectrics and insulators present in the IC.

The three ESD mechanisms described above are known respectively as (1) Human Body Model (HBM), (2) Machine Model (MM), and (3) Charged Device Model (CDM). In this chapter, we will discuss these ESD models in detail, including the methods used to test the sensitivity of ICs to each of these discharge mechanisms.

Although ESD is the result of a static potential in a charged object, the energy dissipation and damage is due to the current in the IC during the discharge. Hence, ESD protection structures need to be capable of sustaining high current levels. ESD sensitivity is generally measured in volts, but this is not really appropriate since the different discharge models can result in vastly different voltages at failure for the same circuit. However, the current at which failure occurs is more or less equivalent in the cases of the HBM and the MM. Even in the CDM, where the failure is due to internal voltage increase, these voltages are the result of the large currents in the IC.

2.2 HUMAN BODY MODEL (HBM)

The HBM is the ESD testing standard, and is defined in the MIL-STD-883C method 3015.7. The discharge waveform of an HBM tester through a 0 Ω load is shown in Figure 2.1. The rise time is approximately 10 ns, and the decay time is around 150 ns. The waveform is obtained by the discharge of a 100 pF capacitor with an initial voltage of 2 kV through a 1.5 kΩ resistor. The HBM can be modeled using the LCR circuit shown in Figure 2.2. C_c is the discharge capacitor and the charging voltage is V_c. L_1 is the parasitic inductance which determines the rise time of the discharge pulse together with the resistor R_1. C_s is the parasitic stray capacitance of R_1 and the interconnect. C_t is the parasitic capacitance of the test board and R_L is the resistance of the load or device under test (DUT) [Chemelli85]. The LCR circuit is easily modeled numerically to obtain accurate waveforms for different values of the elements in Figure 2.2 [Roozendaal90][Verhaege93]. A

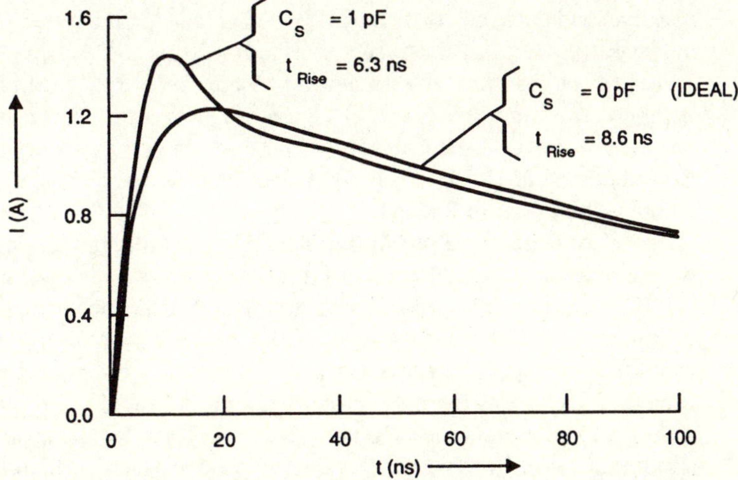

Figure 2.1 HBM discharge waveform through a 0 Ω load. ESD voltage is 2000 V.

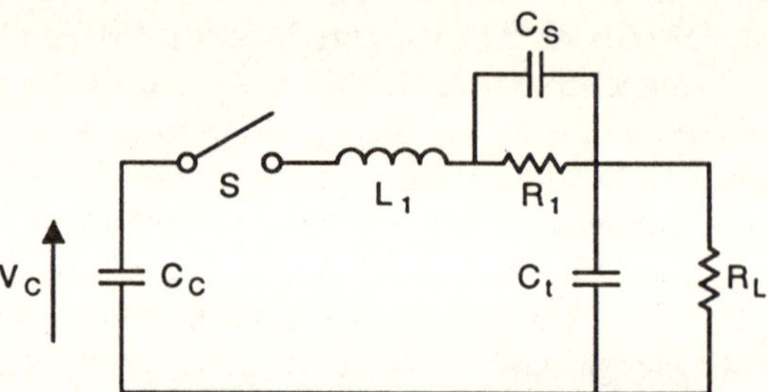

Figure 2.2 Equivalent LCR circuit for modeling HBM and MM discharge waveforms.

simple description of the HBM current waveform can be obtained from a simplified solution for the LCR circuit.

$$I = V_c C_c \frac{\omega_0^2}{\sqrt{a^2 - \omega_0^2}} e^{-\frac{R}{2L_1}t} \sinh(\sqrt{a^2 - \omega_0^2})t \qquad (2.1)$$

with $a = R/2L_1$ and $\omega_0 = 1/\sqrt{L_1 C_s}$, and $a > \omega_0$.

From this equation, an estimate of the risetime is given by

$$t_{rise} = \frac{2L_1}{R} \qquad (2.2)$$

For $t_r = 10$ ns, L_1 is required to be about 7.5 μH.

The effect of the stray capacitance, C_s, is shown in Figure 2.1. The HBM current waveforms for a 2 kV discharge of an ideal circuit with $C_s = 0$ pF and $C_t = 0$ pF, and a circuit with $C_s = 1$ pF and $C_t = 0$ pF, indicate that C_s causes t_r to decrease and the peak current I_p to increase. The waveform is simulated through a short circuit with $R_L = 0$ Ω. The effect of higher test board capacitance, C_t, is significant when the DUT has a high impedance. In that case, I_p decreases by up to 20% for C_t values between 10 pF and 40 pF. The effect of C_t on the HBM waveform is observed when a resistive load between 100 Ω and 1000 Ω is placed in the tester [Roozendaal90]. I_p is observed to decrease and t_r increases from 10 ns to 30 ns as C_t increases with a 1000 Ω load. Calibration with a 500 Ω load is now recommended [Verhaege93] in order to determine the effect of C_t on the discharge waveform. The effect of C_t becomes important when devices with large pin counts are tested. In this case van Roozendaal *et al* [Roozendaal90] showed that I_p and t_r were dependent on the pin combinations used in the test.

Tester parasitics are a serious problem when comparing the ESD failure thresholds of the same device as obtained from different testers. This is a particular concern if the IC manufacturer's tester has a more optimistic behavior than the customer's equipment. Comparisons between different testers [Lin87] [Strauss87] [Roozendaal90] [Verhaege93] have shown that significant differences exist in the output waveform as well as the product ESD failure threshold voltage (ESDV). Figure 2.3 shows the variation in the ESDV for the same IC tested using seven different testers [Roozendaal90]. However, the results are not consistently different and the variations depend on the type of protection circuit and the technology in which the product is

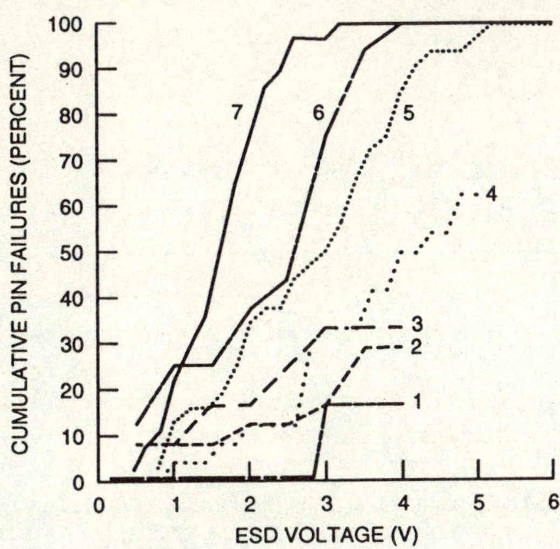

Figure 2.3 Cumulative pin failures vs. ESD voltage for the same product tested with different HBM simulators 1 to 7.

made. The MIL-STD-883C method 3015.7 specifies the rise time of the pulse and the decay time through a zero impedance load at different discharge voltages. However, the high R_1 masks the impedance of the test board and the interconnect. These can only be observed by using varying R_L values. Hence, all the testers in these studies met the MIL-STD specification, but were not identical. Recently, a more stringent standard has been developed which includes calibration through a 500 Ω load. The problem at this point is the high cost of achieving tester compatibility due to the complexities associated with high frequency measurements. The parasitic impedances are a big problem, especially when trying to maintain pin-to-pin compatibility across a test board. IC manufacturers, too, are reluctant to introduce stricter calibration methods because they are expensive (requiring high frequency oscilloscopes, probes etc.) and can result in long periods of equipment downtime. It is important, though, to know the parasitics in the tester and to understand the effect of these parasitics on IC ESD behavior. Furthermore, when comparing data between different testers, the parasitics of both testers should be known. Perhaps these parasitics will eventually be specified as part of the equipment manufacturer's data sheet.

2.3 MACHINE MODEL (MM)

The MM is the standard ESD test method in Japan, and this has driven its usage amongst US and European IC manufacturers. The MM discharge circuit can be defined by the LCR network in Figure 2.2. C_c is defined as 200 pF, while R_1 is required to be 0 Ω. In practical circumstances R_1 will be greater than 0 Ω, and during a discharge the dynamic impedance of the circuit can be much higher than zero. Hence, existing MM standards such as that presented by Philips [Roozendaal90]

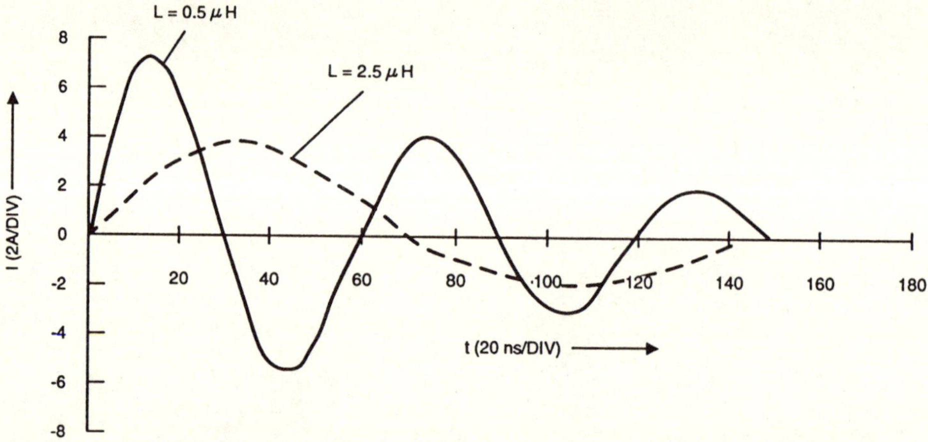

Figure 2.4   Current waveforms from machine model tester. The charging voltage is 400 V. The effect of different inductances, *L*, on the peak current and rise time is also shown.

specify the output current waveform in terms of peak current and oscillating frequency for a given discharge voltage. In this manner, the parasitic L_1 and R_1 are automatically defined for the system. The dependence of the output waveform on L_1 is shown in Figure 2.4 for a 400 V charging voltage. For $L_1 = 0.5$ μH, the peak current, I_p, is about 7 A, with an oscillating frequency of 16 MHz. The effective rise time of the current is about 15 ns. When L_1 is increased to 2.5 μH, I_p drops to 3.7 A for the same charging voltage (400 V) and the oscillating frequency decreases to 7 MHz. The rise time is now about 30 ns. Therefore, the definition of L_1 has a significant role in defining the ESDV of an IC. L_1 is influenced by the test board and the package leads, and for $L_1 < 1$ μH it will be difficult to obtain reproducibility between pins in ICs with large pin counts (> 64 pins). Hence, the MM specification needs to have $L_1 > 1$ μH. The Philips specification requires that $L_1 = 2.5$ μH with $R_1 = 25$ Ω and an oscillating frequency of 7 MHz [Roozendaal90].

Because of the low series resistance in MM testers, the effects of parasitic circuit elements are easily detected. Hence, it is easier to ensure compatibility between testers that have been calibrated through a zero ohm load resistance than for the HBM testers. This was one of the features which led to the development of this test method in Japan, where initial testers had very high L_1 (of the order of 10 μH). The high L_1 produced a double exponential discharge waveform very similar to that of the HBM tester, but variations in test board capacitance and lead inductance were easily detected during calibration. On the negative side, the sensitivity of the MM waveform to the dynamic impedance of the device being tested and the test board, makes pin-to-pin reproducibility for the MM pulse difficult, especially in ICs with large pin counts.

The MM current can be described by [Roozendaal90]:

$$I = V_c C_c \frac{\omega^2}{\sqrt{\omega^2 - a^2}} e^{-\frac{R}{2L_s}t} \sin(\sqrt{\omega^2 - a^2})t \qquad (2.3)$$

with $a = R/2L_1$ and $\omega_0 = 1/\sqrt{L_1 C_s}$, and $a > \omega_0$. When $a < \omega_0$ then we get:

$$I = V_c \sqrt{\frac{C_c}{L_s}}\, e^{-\frac{R}{2L_s}t} \sin\left(\frac{t}{\sqrt{L_s C_c}}\right) \qquad (2.4)$$

$R = (R_1 + R_L)$ is the total resistance of the discharge circuit, including the load resistance, and $t_{max} = 1.5\sqrt{L_1 C_c}$, $t_{rise} = \sqrt{L_1 C_c}$. The amplitude of I is given by $V_c\sqrt{C_c/L_1}$ and the dependence on L_1 can be seen. The oscillating part of the waveform is described by the sinusoidal term, and both C_c and L_1 determine the frequency. The damping is described by the series resistance and the inductance. If R_L does not meet the condition

$$R_L < 2\sqrt{\frac{L_1}{C_c}} - R_1 \qquad (2.5)$$

then the damping term dominates, and I does not oscillate. Typically, for $L_1 = 0.5\ \mu H$, $R_L < 75\ \Omega$ satisfies the condition in Equation 2.5. When $L_1 = 2.5\ \mu H$, $R_L < 200\Omega$ satisfies that condition. In the case that the condition given in Equation 2.5 is not met, then the current waveform becomes that of the HBM, and is described by Equation 2.1.

The MM and the HBM tests are different forms of the same discharge mechanism, that of an externally charged object discharging through the IC. The failure modes are similar, although the severity of the damage varies between the two tests. A comparison between the two methods will be made later in this chapter.

2.4 CHARGED DEVICE MODEL (CDM)

The CDM test method was originally described by Bossard, Chemelli and Unger [Bossard80][Unger81] in order to explain their observed failure mechanisms. Essentially, the HBM and the MM result in similar failure mechanisms based on the energy dissipation in the internal elements of the IC. In contrast, CDM type ESD events result in gate oxide breakdown inside the IC which are related to the internal discharge paths and voltage build-up in the chip. Hence, the CDM is a different type of ESD test and device sensitivity to the CDM cannot be inferred from results of HBM or MM tests. The increased usage of automated manufacturing and testing equipment has led to environments more suitable to CDM type ESD, rather than HBM ESD. In addition, as ICs have become more dense and feature sizes approach $0.5\ \mu m$ with gate oxides of the order of 100 Å, the CDM has become a very relevant test method. A major drawback in developing a reproducible CDM test method is that parasitic elements in the discharge path significantly affect the discharge waveform. Further complications arise in the mechanical complexity required to charge the IC and to avoid damage occurring during the charging process [Shaw86][Enoch86]. In addition, the very fast discharge times, typically less than 1 ns, require expensive oscilloscopes with bandwidths in excess of 2 GHz in order to monitor and calibrate the discharge pulse. Hence, there is as yet no fixed CDM test standard, and available testers are based on either the AT&T tester

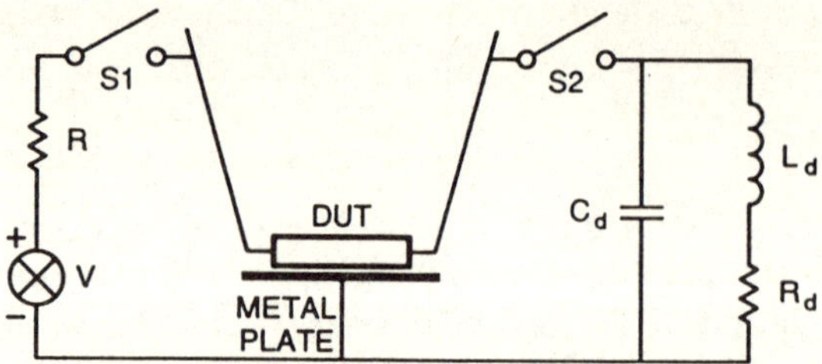

Figure 2.5 Equivalent circuit for a charged device model test.

described by Bossard, Chemelli and Unger [Bossard80] and Renninger *et al* [Renninger89] which is commonly known as the *non-socketed test method*, or the direct charging method which is also known as the *socketed test*.

The original CDM was based on the direct charging method. Figure 2.5 shows the simplified equivalent circuit of the tester. The IC is placed with its pins upwards on a metal plate. The charge is applied via the switch S1 and the resistor, R, directly to one of the pins. S1 is then opened. The capacitance between the silicon, the package and the metal plate stores the charge until the switch S2 is closed. The discharge then takes place through the discharge probe with equivalent capacitance, C_d, inductance, L_d, and resistance, R_d. L_d is a crucial parameter of the test system and needs to be minimized in order to maintain the fast pulses of < 1 ns, typical of CDM type discharges. Figure 2.6 shows a CDM current waveform as a function of time [Lin92] obtained experimentally. The charging voltage was 1100 V and the peak current was greater than 6 A with a rise time of $\approx$ 0.5 ns. The calculated waveform also shown in Figure 2.6 is based on the equation

$$I(t) = \mu CDE \int_E^{E_0} \alpha \, dE' \tag{2.6}$$

C is the parallel-plate capacitance, D is the gap distance, E is the electric field and α is an ionization coefficient [Lin92]. μ is the electron mobility in the gap. The close fit with experimental data is made with $\mu = 105$ cm^2/V s, $C = 2.8$ pF, and

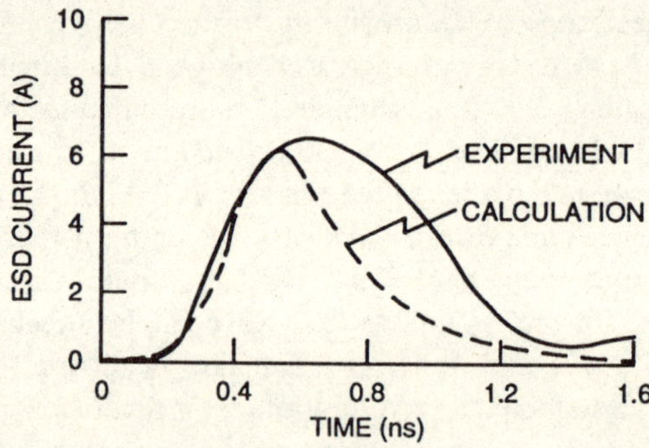

Figure 2.6 Current waveform for a CDM type discharge for a charging voltage of 1100 V (after [Lin92]). The experimental curve (solid) is compared with the calculated curve (dashed) using Equation 2.6.

$D = 0.01$ cm. E is determined from the charging voltage and α in cm^{-1} is given by the equation

$$\alpha = 6.08 \times 10^3 \exp\left(-\frac{1.88 \times 10^5}{E}\right) \tag{2.7}$$

If $L_d > 0.1$ μH, then the discharge waveform looks very similar to that of the MM tester. In practical applications, $L_d = 50$ nH can be achieved. Since C_d also influences the discharge waveform, it needs to be minimized [Fukuda88]. C_d values in the proximity of 2 pF to 6 pF have been obtained [Renninger91]. Under all conditions the path resistance, R_d, must be of the order of 1 Ω. This is a very low value, and requires some degree of complexity in the design of the discharge probe [Renninger89].

Because of some of the problems described above, variations on the CDM based on indirect charging have been developed [Fukuda88][Renninger91]. A major advantage of this approach, which is also called the *field-induced method*, is that the entire package is charged by induction in the presence of an electric field. The metal plate in Figure 2.6 is replaced by a dielectric material such as plastic or wood which is in contact with a metal charging electrode. The capacitance of the dielectric has a significant influence on the amount of charge in the IC at a given threshold voltage. It will, therefore, influence the failure threshold voltage level for the IC. The discharge circuit is similar to that for the direct charging method described above. A similar test method has been developed for wafer-level CDM testing as described by Fukuda and Kato [Fukuda88].

2.5 COMPARISONS BETWEEN THE TEST METHODS

Figure 2.7 shows a comparison of the frequency spectrum for the discharge of the HBM, MM and CDM into a 10 Ω load. For the HBM, $C_c = 100$ pF, $L_1 = 7.5$ μH, and $R_1 = 1500$ Ω. For the MM, $C_c = 200$ pF, $L_1 = 2.5$ μH, and $R_1 = 25$ Ω. For the CDM, $C = 5$ pF, $L_1 < 10$ nH, and $R_1 < 10$ Ω. It is seen that the HBM has a bandwidth of 2.1 MHz with a center frequency of 500 kHz, while the MM has a bandwidth of 12 MHz with a center frequency of 7 MHz. The CDM on the other hand has a bandwidth of over 1 GHz with a center frequency in excess of 600 MHz. At 100 MHz, the skin effect in an aluminum strip requires 8 μm, which is larger than the thickness of any metal in the IC. At 10 GHz, the skin effect requires only 0.8 μm, and can begin to influence the conductivity of the metal lines. However, 10 GHz is still much higher than the expected bandwidth of the CDM discharge which implies that the metal resistance will be close to its DC value.

Failure modes for the HBM and the MM test methods are typically found in the diffusion regions of the protection circuits [Avery87][Fukuda88][Kitamura89] [Renninger89][Roozendaal90]. However, for the CDM, the duration of the pulse is less than 1 ns, which can be less than the time required to trigger the protection circuits. Hence, CDM failures are usually gate oxide damage either at the pad, or in some

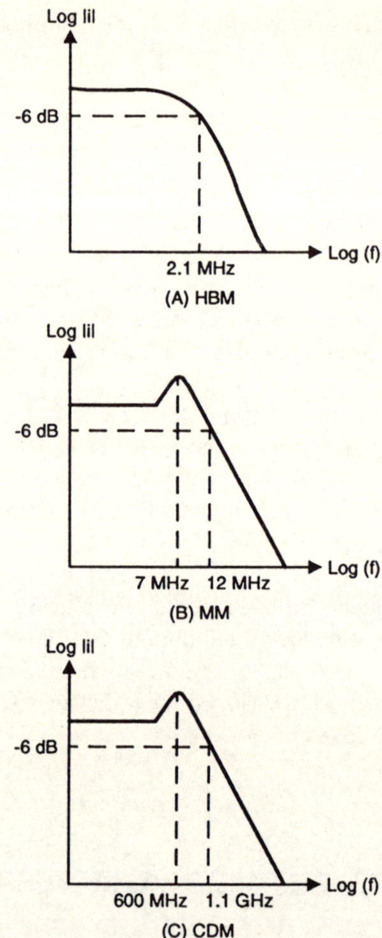

Figure 2.7   Frequency spectrum of the current waveform calculated for (A) human body model (HBM), (B) machine model (MM), (C) charged device model (CDM).

cases internally in the circuit [Unger81][Fukuda86][Avery87][Fukuda88] [Maloney88][Kitamura89]. The most common location for gate oxide damage is in the pMOS transistor of the input buffer [Chaine94][Gieser94]. Fukuda, Ishiguro and Takahara [Fukuda86] showed that, with decreasing gate oxide thickness from 500 Å to 300 Å, the sensitivity to the CDM test method was increased by almost 70%. In the same range of oxide thickness the sensitivity to the HBM increased by only 30%. It was also shown that the resistance of the protection circuit increased the robustness to CDM type pulses. The robustness to HBM type stress was also increased by a similar amount (approximately 2×). Damage in the intermetallic oxide or passivation layer due to CDM pulses has also been observed [Renninger89]. One of the issues raised with CDM type damage is that of cold healing, whereby the damage anneals with time at room temperature [Verhaege94]. It is not certain whether such effects can lead to a latent reliability hazard, but the possibility must certainly be considered when evaluating the effect of CDM damage.

Reproducibility for the HBM and the MM are generally made more difficult by the conventional measurement of ESD failure thresholds in terms of voltage. As far as the device is concerned, the high resistance in the HBM makes it a current source, and it would be more appropriate to measure failure in terms of peak current rather than voltage. To all intents and purposes, the MM is a current source as well, although the use of large resistors (>100 Ω) in the protection circuit network can change that. However, correlation between the HBM and the MM failure thresholds is possible if the threshold levels are measured in current rather than voltage. The failure current of a particular circuit for both test methods is comparable, provided the circuit has been properly designed for optimized current flow and power dissipation. The failure voltage, on the other hand, may differ by a factor of 10 with the HBM having a higher ESDV than the MM. In general, if the HBM and MM failure currents differ by more than 50%, or if the failure modes differ widely, then the design of the protection circuit probably favors one method over the other. This is usually due to the impedance of the protection circuit [Hyde89].

The reproducibility of the HBM test has been reported in detail [Chemelli85] [Strauss87][Lin87][Roozendaal90][Verhaege93]. It is reported that the main cause of variations in results is the influence of the parasitic inductances and the stray capacitances on the discharge waveform. Recent changes to HBM characterization techniques will minimize these differences. In particular, calibration of the waveform through a zero ohm load, as well as a 500 Ω load, will enable the parasitics to be properly identified and compensated.

The reproducibility of MM results is strongly dependent on the choice of inductance. Testers with different inductances will not have similar results. However, testers which meet the same waveform specification for current as described above will show comparable results [Roozendaal90]. Comparisons between the two CDM methods have shown that correlations are difficult both between the test methods and even between different testers with the same test method [Verhaege94][Gieser94][Chaine94]. However, both test methods have been able to locate and reproduce real-life CDM-type failure modes [Chaine94].

2.6 TRANSMISSION LINE TEST METHOD

The high current, short duration pulse can also be easily reproduced using a charged transmission (coaxial) line [Brown72][Khurana85][Maloney85]. The advantage of the transmission line pulse (TLP) tester is that a constant current pulse is generated which enables the behavior of the protection structure under high current conditions to be studied. The double exponential and oscillating pulses of standard ESD testers make it difficult to determine how the protection structure operates. Hence, the TLP tester has become popular with many ESD protection circuit designers and researchers into protection circuit operation and physics [Pierce88][Bridgewood88] [Chen88][Polgreen89][Amerasekera90][Abderhalden91][Diaz92]. The relationship between the TLP and the HBM pulse was shown by Pierce *et al* [Pierce88] leading to the possibility that the TLP may be used as a universal ESD test method at some future date. The HBM pulse was shown to be equivalent to a constant current pulse of approximately 100 ns duration.

The TLP tester configuration has the advantage that it is easily designed to avoid sensitivity to internal parasitic elements. Hence, we obtain reproducible testing which can be correlated to HBM or MM type discharges, or any other discharge waveform which one may decide upon (except the CDM which is a different type of stress). Figure 2.8 shows a circuit schematic for a TLP generator. The transmission line is a 50 Ω coaxial cable which is charged by means of the DC high voltage generator. The cable is terminated in 50 Ω at either end to ensure that no reflections occur. The discharge voltage is converted into a current pulse by means of a resistor, R. For optimum termination and minimum pulse distortion due to reflections, R needs to be at least 500 Ω [Abderhalden91]. R also ensures that the resistance of the protection circuit being tested does not cause variations in the current and the pulse waveform. The pulse width is determined by the length of the cable, and is typically about 10 ns/m. That is, a 10 m cable would be needed for a 100 ns pulse. A typical current and voltage waveform for a protection circuit stressed with the TLP tester is shown in Figure 2.9. The rise time of the pulse can be very short ($<$ 1 ns) but for reproducibility and calibration purposes which rely on the oscilloscope bandwidth, it is best to make the rise time at least 1 ns.

The voltage waveform shows the avalanche breakdown and triggering of a lateral bipolar transistor which is used in MOS protection circuits (Chapter 3). The initial rise in the voltage occurs while the lateral bipolar is turning on. After turn-on, the voltage drops to the snapback holding voltage. It must be noted that the voltage peak before snapback is also influenced by the charging capacitance of the system and the parasitic inductance, and is not an accurate representation of the voltage across the device as injection current levels are increased ($>$100 mA). If the current pulse is sufficiently large, the heat dissipated in the lateral bipolar transistor eventually causes thermal or second breakdown to occur and the voltage collapses. By monitoring the time taken before second breakdown occurs, the voltage at this point and the magnitude of the current, the power required to cause failure can be determined. The impact of the pulse duration can be easily obtained by varying the pulse duration and determining the variations in the power required to cause failure [Wunsch68][Brown72][Horgan87][Pierce88][Amerasekera91][Diaz92]. Hence, useful information about the heat dissipation and high current capability of the device and the technology can be gained. The analysis can lead to improvements

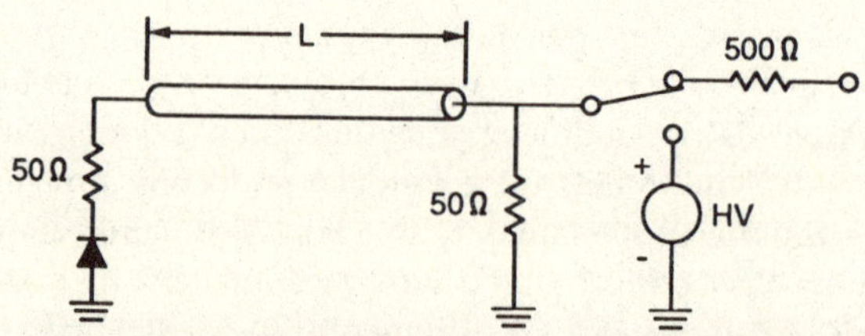

Figure 2.8   Transmission line pulse generator.

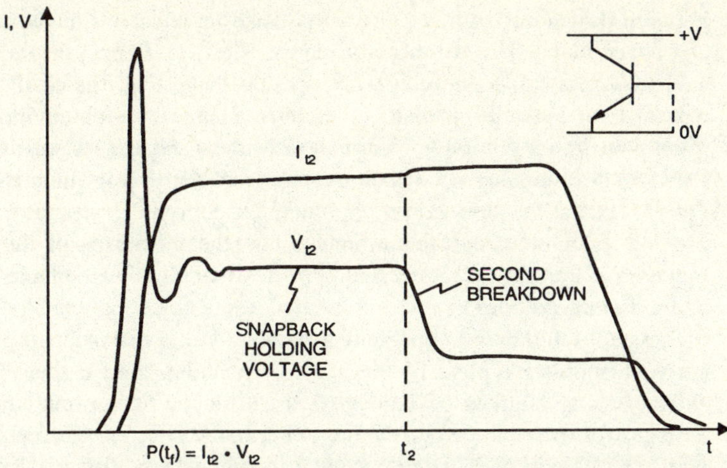

$$P(t_f) = I_{t2} \cdot V_{t2}$$

Figure 2.9   Current and voltage waveforms obtained with a tranmsmission line pulse generator. The pulse is applied to a parasitic bipolar device (inset). At second breakdown the power-to-failure is defined by $P(t_f) = I_{t2} \cdot V_{t2}$.

in the development of ESD protection circuits and even changes in the technology to improve the important parameters influencing heat dissipation, thermal breakdown and failure [Amerasekera91].

2.7 OTHER TEST METHODS

The test methods described above are the principal ones in use in the IC environment. Other test methods have been developed for evaluating the ESD sensitivity of ICs to be used in specific applications such as automotive or military applications [Fisher89][Blankenagel89][Lee88][Sullivan85][AEC94]. Circuits which are required to operate in hostile environments such as in automobiles need to be able to withstand much higher ESD type stress conditions. These stress conditions are not readily reproducible with standard HBM/MM type testers, and variations which use diff-erent combinations of capacitance and series resistance have been developed. It must also be noted that circuits used in these applications usually have more robust protection circuits than those in standard ICs.

2.8 FAILURE CRITERIA

The nature of ESD damage in ICs is such that the selection of the failure criteria can have a significant influence on the ESD failure threshold. Not all instances of ESD stress result in a short circuit at the pin under test. Variations in the severity of the damage can result in large differences in the post-stress leakages measured [Amerasekera90][Ohtani90][Amerasekera92]. Furthermore, the result can be an increase in the leakage current either at the stressed pin or between other pins, especially between the supply pins. Changes may also be observed in the access times which can affect the behavior of fast ICs such as SRAMs. Hence, the stressed IC should be subjected to a full-functional test to ensure that it still meets all the product specifications for leakage current and performance. However, in most cases, it is not possible to run a complete electrical test after each ESD stress. Commercially available ESD testers are usually not capable of performing more than simple leakage current tests at the stressed pin, although more complex (and expensive) testers will provide leakage current measurements for other pins as well. For quick evaluation, the leakage current for the stressed pin at a specified bias voltage is usually selected as the failure criterion.

Figure 2.10 shows a series of typical $I\text{–}V$ curves measured after ESD stress. Curve 1 shows a typical $I\text{–}V$ curve for an undamaged device. Curves 2, 3 and 4 show varying degrees of damage after different ESD stress. Typical monitoring voltages for the leakage current are either 0.4 V or 5.5 V (or 3.6 V for 3.3 V parts), while the choice of leakage current can range from 100 nA to 10 μA. It is also possible to select a percentage change in leakage current (10% to 50%). From Figure 2.10 we see that for Curve 3 the choice of a 1 μA leakage current will signal failure at 5.5 V, but 100 nA at 0.4 V will not show a failure. Similarly Curve 2 will pass most failure criteria, but in reality the device has been damaged when compared to Curve 1. There is no certainty that such a pin will show latent effects and be a reliability hazard in device operation, but it has definitely been damaged. It is indeed possible that the damage may manifest itself as a change in the access time of a fast SRAM, or a malfunctioning of differential amplifier type circuits. Therefore, one must be careful when comparing ESD failure thresholds obtained with different failure criteria, and with the choice of the failure criterion itself. It is recommended that during ESD protection circuit evaluation and design development, all changes in leakage current are monitored and signalled as the onset of

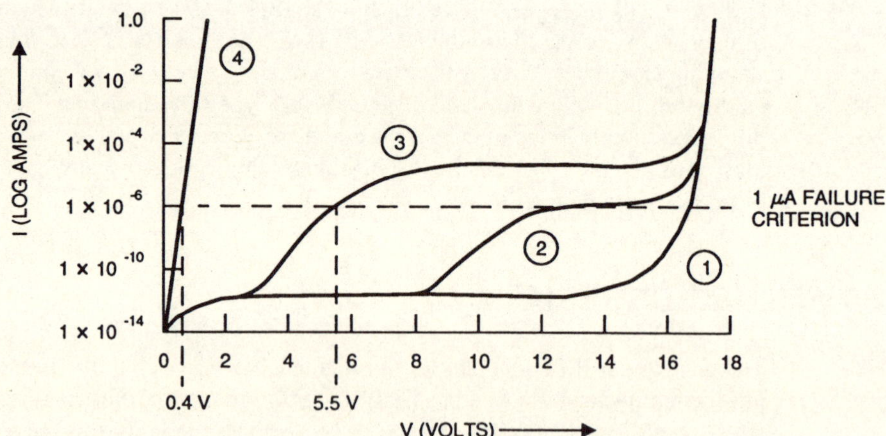

Figure 2.10   Leakage curves measured after ESD stress across an nMOS transistor. Curves 1 to 4 indicate increasing levels of damage severity.

failure. It has been shown that the distribution of the leakage current at a given stress level is linked to the distribution of the ESD failure threshold [Amerasekera90], and a monitor of the leakage current can be used to characterize the ESD behavior of an IC.

The increase in standby current, or the leakage current between the V_{cc} and V_{ss} pins, is an issue which is important in certain chips and technologies. For example, floating substrate ICs are especially susceptible to damage between V_{cc} and V_{ss} after an ESD stress. Similarly, protection circuits which require the stress current to be conducted via a V_{cc} to V_{ss} path are also likely to result in V_{cc} to V_{ss} damage and increased leakage current between the supply and ground, rather than leakage at the stressed pin itself. These examples emphasize the need for a careful consideration of the type of IC being tested when selecting the failure criteria.

2.9 TEST PROCEDURES

Most semiconductor and system manufacturers have in-house test procedures which are adapted to meet their own requirements. For the most part, they follow the general test procedure defined by the MIL-STD-883C method 3015. The procedure specifies the calibration of the tester, the number of samples to be tested and the pin combinations to be tested. MIL-STD-883C method 3015 only deals with the HBM tester. However, the procedure can be applied to other test methods as well.

The selection of pin combinations is a difficult issue when dealing with ICs with large pin counts (>64). Each pin is stressed with either the substrate pin, V_{ss}, grounded or the supply pin, V_{cc}, grounded. All other pins are left floating. All the pins are also required to be stressed against every other pin in turn, but the large number of combinations make this unfeasible. A reasonable judgement must be made of which pins will provide the worst-case conditions under pin-to-pin testing in order to reduce testing time. These conditions will vary according to the type of silicon (n-substrate or p-substrate), the type of IC (e.g. NMOS, CMOS, bipolar) and the circuit design itself. As a rule of thumb, the pins furthest apart from each other (i.e. diagonally opposite), and the pins adjacent to the stressed pin, would provide an indication of the worst-case performance. However, it may be necessary to check the layout of the IC and the bonding of the package to ensure that these are indeed the pins with the largest and smallest resistances to the stressed pin. A preferred approach to this issue is to stress the pin with all the other non-supply pins grounded. The effect is the same as a pin-to-pin test, but it provides a number of return paths for the stress current which will reduce the severity of the test compared to a direct pin-to-pin test.

The general requirement is that the stress is performed with three positive and three negative pulses spaced at intervals of 1 second to allow a cooling period for the device under test. Three pulses have been selected to ensure that any possible cumulative degradation as a result of repeated pulsing will manifest itself.

For qualification purposes, devices are stressed at the required pass voltage with fresh devices being used for each stress condition. A large number of devices are required for such a test, and for the purpose of characterization of ESD behavior (engineering purposes) it is more practical (and cheaper) to step stress the device until failure. The disadvantage with step stressing is that the cumulative degradation at each voltage step may influence the final failure threshold voltage, which could actually be higher or lower than the single stress value. Hence, step stressing may not provide accurate values of the absolute failure threshold but it does provide a quick way of estimating the performance levels of a protection circuit or device, or for determining trends. An additional feature of step stressing is that a failure window may exist for the ESD threshold voltage of a given protection circuit [Duvvury88]. At ESD voltages outside this window, both above and below, the protection circuit works, but not in the intermediate region. Single voltage testing can miss these windows and result in an inaccurate evaluation of the ESD capability of the circuit.

Testing specifications usually require circuits to have a minimum pass threshold voltage of $\pm$ 2000 V HBM stress on all pins, with respect to all supply and ground pins, independently, and with respect to other pins. However, ESD pass voltages of 1000 V HBM are still acceptable for more complex ICs such as microprocessors and microcontrollers as well as large/fast (>1 Mb) SRAMs and DRAMs. More aggressive manufacturers have developed 4000 V ESD robust ICs, even for advanced circuits in state-of-the-art technologies, creating a target level which other manufacturers are urged to match. There is a good reason for this – US military specifications require that all products with less than 4 kV ESD thresholds are labeled ESD sensitive and are subjected to special handling procedures and restrictions which may add to the cost of manufacturing. Many commercial specifications have similar standards.

The choice of 2 kV and 4 kV as threshold levels for ESD qualification have no specific meaning in themselves. However, there is evidence that products with ESD thresholds <2 kV are prone to high fallout during assembly, packaging and handling, leading to lower product yields. In the manufacture of large ICs the product yield can be of the order of 30% to 50% and a further 10% fallout of electrically good chips on top of that is costly to the IC manufacturer. Fallout decreases as ESD levels are increased to 4 kV. Increasing the ESD threshold beyond 4 kV does not significantly decrease the fallout. Therefore, the additional cost required to achieve ESD thresholds >4 kV is not worthwhile for ICs to be used in computer or telecommunication applications, which is the major share of the market for state-of-the-art ICs. The automotive industry, on the other hand, does require even higher ESD levels in some cases, ranging up to 10 kV or more, because of the hostile environments in which these parts are required to operate.

Because of the higher current level for the MM discharge voltage compared to the HBM, the pass thresholds for commercial ICs are typically 200 V. However, as in the case for the HBM requirement a target specification of 400 V is set by more aggressive manufacturers, while ICs using new technologies and complex circuit functions can be passed at 100 V levels. In addition, it must be emphasized that the ESD failure threshold voltage is a sensitive function of the inductance of the test system. A 200 V stress using a tester with an inductance of 2.5 μH is less severe than a 200 V stress using a tester with a 0.5 μH inductance and may lead to confusing MM results when analyzing data using different test systems.

Typical CDM pass levels are 1 kV. However, the absence of either an industry standard or a fixed test method, coupled with a limited understanding of the test method itself, has resulted in a very arbitrary setting of the pass thresholds.

REFERENCES

[Abderhalden91] J. Abderhalden, *Untersuchungen zur Optimierung von Schutzstrukturen gegen elektrostatische Entladungen in integrierten CMOS-Schaltungen*, PhD Thesis, Eidgenossische Technische Hochschule (ETH), Zurich, 1991.

[AEC94] Automotive Electronics Council 'Stress Test Qualification for Automotive-Grade Integrated Circuits', CDF-AEC-Q100, June 1994.

[Amerasekera90] E. A. Amerasekera, L. J. van Roozendaal, J. Abderhalden, J. J. P. Bruines, L. Sevat, 'An Analysis of Low Voltage ESD Damage in Advanced CMOS Processes', in *Proc. 12th EOS/ESD Symposium*, p. 143–150, 1990.

[Amerasekera91] A. Amerasekera, L. van Roozendaal, J. Bruines, F. Kuper, 'Characterization and Modeling of Second Breakdown in NMOST's for the Extraction of ESD-related Process and Design Parameters', *IEEE Trans. Elec. Dev.*, ED-38, p. 2161-2168, 1991.

[Amerasekera92] E. A. Amerasekera, W. van den Abeelen, L. J. van Roozendal, M. Hannemann, P. J. Schofield, 'ESD Failure Modes: Characteristics, Mechanisms and Process Influences', *IEEE Trans. Elec. Dev.*, ED-39, 2, 1992.

[Avery87] L. Avery, 'Charged Device Model Testing: Trying to Duplicate Reality', in *Proc. 9th EOS/ESD Symposium*, p. 88–92, 1987.

[Blankenagel89] J. Blankenagel, L. Lockwood, 'A High Voltage Pulse Generator for ESD Simulation', in *Proc. 11th EOS/ESD Symposium*, p. 50–54, 1989.

[Bossard80] P. Bossard, R. Chemelli, B. Unger, 'ESD Damage from Triboelectrically Charged IC Pins', in *Proc. 2nd EOS/ESD Symposium*, p. 17–22, 1980.

[Bridgewood88] M. Bridgewood, Y. Fu, 'A Comparison of Threshold Damage Processes in Thick Field Oxide Protection Devices Following Square Pulse and Human Body Model Injection', in *Proc. 10th EOS/ESD Symposium*, p. 129–136, 1988.

[Brown72] W. D. Brown, 'Semiconductor Device Degradation by High Amplitude Current Pulses', *IEEE Trans. Nucl. Sci.*, NS-19, p. 68–75, Dec. 1972.

[Chaine94] M. Chaine, C. T. Liong, H. F. San, 'A Correlation Study between Different Types of CDM Testers and Real Manufacturing In-Line Leakage Failures', in *Proc. 16th EOS/ESD Symposium*, p. 63–68, 1994.

[Chemelli85] R. Chemelli, L. De Chiaro, 'The Characterization and Control of Leading Edge Transients from Human Body Model ESD Simulators', in *Proc. 7th EOS/ESD Symposium*, p. 155–162, 1985.

[Chen88] K.-L. Chen, 'Effect of Interconnect Process and Snapback Voltage on the ESD Failure Threshold of NMOS Transistors', in *Proc. 10th EOS/ESD Symposium*, p. 212–219, 1988.

[Diaz92] C. Diaz, S. Kang, C. Duvvury, L. Wagner, 'Electrical Overstress (EOS) Power Profiles: A Guideline to Qualify EOS Hardness of Semiconductor Devices', in *Proc. 14th EOS/ESD Symposium*, p. 88–94, 1992.

[Duvvury88] C. Duvvury, R. Rountree, O. Adams, 'Internal Chip ESD Phenomena beyond the Protection Circuit', in *Proc. 26th IRPS*, p. 19–25, 1988.

[Enoch86]   R. D. Enoch, R. N. Shaw, 'An Experimental Validation of the Field Induced ESD Model', in *Proc. 8th EOS/ESD Symposium*, p. 224–231, 1986.

[Fisher89]   R. Fisher, 'A Severe Human Body ESD Model for Safety and High Reliability System Qualification Testing', in *Proc. 11th EOS/ESD Symposium*, p. 55–58, 1989.

[Fukuda86]   Y. Fukuda, S. Ishiguro, M. Takahara, 'ESD Protection Network Evaluation by HBM and CPM (Charged Package Method)', in *Proc. 8th EOS/ESD Symposium*, p. 193–199, 1986.

[Fukuda88]   Y. Fukuda, K. Kato, 'VLSI ESD Phenomenon and Protection', in *Proc. 10th EOS/ESD Symposium*, p. 228–234, 1988.

[Gieser94]   H. Gieser, P. Egger, 'Influence of Tester Parasitics on Charged Device Model Failure Thresholds', in *Proc. 16th EOS/ESD Symposium*, p. 69–84, 1994.

[Horgan87]   E. Horgan, 'Advanced Semiconductor Device EOS Test and Modeling Methods', in *Proc. 9th EOS/ESD Symposium*, p. 174–178, 1987.

[Hyde89]   P. Hyde, H. Domingos, 'Analysis of ESD Protection on Vendor Parts', in *Proc. 11th EOS/ESD Symposium*, p. 72–77, 1989.

[Khurana85]   N. Khurana, T. Maloney, W. Yeh, 'ESD on CHMOS Devices–Equivalent Circuits, Physical Models and Failure Mechanisms', in *Proc. 23rd IRPS*, p. 212–223, 1985.

[Kitamura89]   Y. Kitamura, H. Kitamura, K. Nakanishi, Y. Shibuya, 'Breakdown of Thin Gate-Oxide by Application of Nanosecond Pulse as ESD Test', in *Proc. Int. Test and Failure Anal. Symp. (ISTFA)*, p. 193–199, 1989.

[Lee88]   T. Lee, G. Thome, T. Guthrie, 'Discrete Semiconductor Testing to 40 kV', in *Proc. 10th EOS/ESD Symposium*, p. 119–123, 1988.

[Lin87]   D. L. Lin, M. Strauss, T. Welsher, 'On the Validity of ESD Threshold Data Obtained Using Commercial Human Body Model Simulators', in *Proc. 25th IRPS*, p. 77–87, 1987.

[Lin92]   D. L. Lin, T. L. Welsher, 'From Lightning to Charged-Device Model Electrostatic Discharges', in *Proc. 14th EOS/ESD Symposium*, p. 68–75, 1992.

[Maloney85]   T. Maloney, N. Khurana, 'Transmission Line Pulsing Techniques for Circuit Modeling of ESD Phenomena', in *Proc. 8th EOS/ESD Symposium*, p. 49–54, 1985.

[Maloney88]   T. J. Maloney, 'Designing MOS Inputs and Outputs to Avoid Oxide Failure in the Charged Device Model', in *Proc. 10th EOS/ESD Symposium*, p. 220–227, 1988.

[Moss82]   R. Y. Moss, 'Caution–Electrostatic Discharge at Work', *IEEE Trans. Comp. Hyb. and Man.*, CHMT-5, p. 512–515, 1982.

[Ohtani90]   S. Ohtani, M. Yoshida, 'Model of Leakage Current in LDD Output MOSFET Due to Low-Level ESD Stress', in *Proc. 12th EOS/ESD Symposium*, p. 177–181, 1990.

[Pierce88]   D. G. Pierce, W. Shiley, B. D. Mulcahy, K. E. Wagner, M. Wunder, 'Electrical Overstress Testing of a 256K UVEPROM to Rectangular and Double Exponential Pulses', in *Proc. 10th EOS/ESD Symposium*, p. 137–146, 1988.

[Polgreen89]   T. Polgreen, A. Chatterjee, 'Improving the ESD Failure Threshold of Silicided nMOS Output Transistors by Ensuring Uniform Current Flow', in *Proc. 11th EOS/ESD Symposium*, p. 167–174, 1989.

[Renninger89] R. G. Renninger, M.-C. Jon, D. L. Lin, T. Diep, T. L. Welsher, 'A Field-induced Charged Device Model Simulator', in *Proc. 11th EOS/ESD Symposium*, p. 59–71, 1989.

[Renninger91]   R. Renninger, 'Mechanisms of Charged Device Model Electrostatic Discharges', in *Proc. 13th EOS/ESD Symposium*, p. 127–143, 1991.

[Roozendaal90]   L. J. van Roozendaal, E. A. Amerasekera, P. Bos, W. Baelde, F. Bontekoe, P. Kersten, E. Korma, P. Rommers, P. Krys, U. Weber, P. Ashby, 'Standard ESD Testing', in *Proc. 12th EOS/ESD Symposium*, p. 119–130, 1990.

[Shaw86]   R. N. Shaw, 'A Programmable Equipment for ESD Testing to the Charged Device Model', in *Proc. 8th EOS/ESD Symposium*, p. 232–237, 1986.

[Strauss87]   M. Strauss, D. Lin, T. Welsher, 'Variations in Failure Modes and Cumulative Effects Produced by Commercial Human Body Model ESD Simulators', in *Proc. 9th EOS/ESD Symposium*, p. 59–63, 1987.

[Sullivan85]   S. Sullivan, D. Underwood, 'The Automobile Environment: Its Effects on Human Body ESD Model', in *Proc. 7th EOS/ESD Symposium*, p. 103–106, 1985.

[Unger81]   B. Unger, 'Electrostatic Discharge Failures of Semiconductor Devices', in *Proc. 19th IRPS*, p. 193–199, 1981.

[Verhaege93]   K. Verhaege, P. Roussel, G. Groeseneken, H. Maes, H. Gieser, C. Russ, P. Egger, X. Guggenmos, F. Kuper, 'Analysis of HBM ESD Testers and Specifications Using 4th Order Lumped Element Model', in *Proc. 15th EOS/ESD Symposium*, p. 129–138, 1993.

[Verhaege94]   K. Verhaege, G. Groeseneken, H. Maes, 'Influence of Tester, Test Method and Device Type on CDM Testing', in *Proc. 16th EOS/ESD Symposium*, p. 49–62, 1994.

[Wunsch68]   D. C. Wunsch, R. R. Bell, 'Determination of Threshold Failure Levels of Semiconductor Diodes and Transistors Due to Pulse Voltages', *IEEE Trans. Nucl. Sci.*, NS-15, p. 244, Dec. 1968.

3

PHYSICS AND OPERATION OF ESD PROTECTION CIRCUIT ELEMENTS

3.1 INTRODUCTION

Devices subjected to high current and high voltage stress are operated well outside their normal working ranges and their modes of operation are very different under these conditions compared to normal operating conditions. An understanding of the high current device behavior is essential in analyzing the phenomena taking place in the IC during an ESD or high current stress event. In this chapter we will look at the behavior of the main circuit elements used in ESD protection circuits. We will discuss the physics of these devices under high current conditions and the main parameters governing their performance.

The principles of high current operation are discussed in books on power semiconductor devices [Ghandhi77]. For the most part, these principles remain the same under ESD conditions. The important devices to be considered are *pn* diodes under both forward and reverse biased conditions, the nMOS transistor, the *npn* transistor, and semiconductor resistor elements. In addition, an important ESD circuit element is the Silicon-Controlled-Rectifier (SCR) and we will discuss its operation. When considering the high current behavior of these devices, we will focus on those conditions directly related to the behavior under ESD conditions.

A basic understanding of semiconductor device physics is assumed throughout this chapter. The background for much of the analysis presented here may be found in books by Ghandhi [Ghandhi77], Sze [Sze81] and Muller and Kamins [Muller86]. The reader is urged to refer to these texts for more detailed information on the contents of this chapter.

3.2 RESISTORS

Resistor structures used in ESD protection circuits can be built either in the silicon using n-type or p-type diffusions, or above the silicon using deposited polysilicon. Polysilicon resistors built above the silicon are not recommended when high current conduction is required because of the poor heat dissipation capability of these elements which are totally isolated from the substrate by silicon dioxide. In general, n-type resistors are preferred in processes which use p-substrates and p-type resistors with n-substrates.

The conductivity of a semiconductor is given by

$$\sigma = nq\mu_n + pq\mu_p \tag{3.1}$$

where μ_n and μ_p are the electron and hole mobilities and n and p are the electron and hole concentrations. $q = 1.602 \times 10^{-19}$ C is the electronic charge, n and p are given by the equations

$$n = n_i \exp\left[\frac{q(\psi - \phi)}{kT}\right] \tag{3.2}$$

$$p = n_i \exp\left[\frac{q(\phi - \psi)}{kT}\right] \tag{3.3}$$

where ϕ and ψ are the potentials corresponding to the Fermi energy, E_F, and the Fermi energy for the intrinsic semiconductor, E_i, respectively. Hence, $\phi \equiv -E_F/q$ and $\psi \equiv -E_i/q$. n_i is the intrinsic carrier concentration given by

$$n_i = \sqrt{N_C N_V} \exp\left(\frac{-E_g}{2kT}\right) \tag{3.4}$$

where N_C and N_V are the effective density of states in the conduction and valence bands, respectively, Eg is the bandgap. For silicon, we get

$$n_i = 1.69 \times 10^{19} \left(\frac{T}{300}\right)^{3/2} \exp\left(\frac{-E_g}{2kT}\right) \tag{3.5}$$

In n-type resistors at low injection current levels, the current density is given by

$$J = J_n = N_B q \mu_n E \tag{3.6}$$

$$= N_B q v_d \tag{3.7}$$

since the hole current is negligible. N_B is the background doping concentration and E is the electric field. $v_d = \mu E$ is the drift velocity of carriers. As the voltage is increased E increases and so does J and the resistor is 'ohmic' i.e. it shows a linear dependence between current and voltage. As the voltage is raised further the field increases and so does J in accordance with Equation (3.6) until at $E = 10^4$ V/cm the electron drift velocity saturates at $v_s \approx 10^7$ cm/s. Further increasing the voltage serves only to increase the electric field with no increase in J and

$$J = J_{sat} = N_B q v_s \tag{3.8}$$

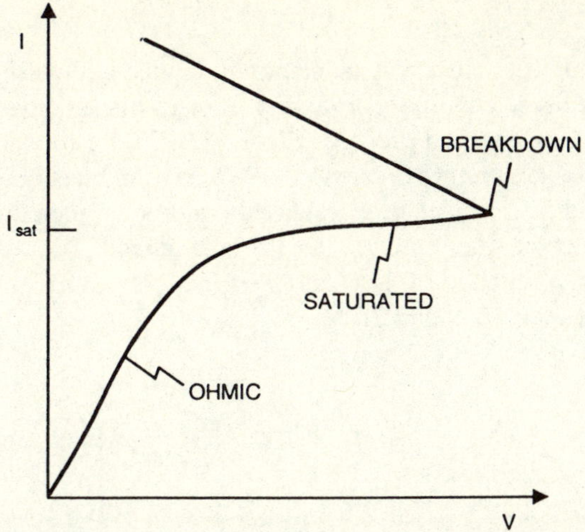

Figure 3.1 The high current I–V curve for an n-type diffused resistor. The ohmic and saturation regions are indicated.

J_{sat} is a function of the doping concentration and will not be observed for either low N_B ($< 10^{14}$/cm^3) or very high doping levels when $N_B \approx 10^{20}$/cm^3 [Hower70][Caru-kso74][Amerasekera93]. For $N_B = 10^{20}$/cm^3, $J_{sat} \approx 10^8$ A/cm^2 which gives I_{sat} for a 20 μm wide resistor of $\approx$10 A. However, n-well resistors with $N_B \approx 10^{17}$/cm^3 will have $J_{sat} \approx 10^5$ A/cm^2 which translates to a current of about 10 mA in a 20 μm wide resistor. An experimental high current I–V curve for a 20 μm wide n-well resistor is shown in Figure 3.1. As the voltage is increased even further in the saturation region, the E-field eventually reaches the impact ionization threshold and holes are generated. When the generated hole current becomes large enough to contribute to the total current the voltage begins to decrease and a negative resistance or *snapback* characteristic is observed as shown in Figure 3.1. A rough calculation of the maximum voltage across the resistor may be done by assuming that impact ionization begins at $\approx$150 kV/cm. Then the maximum voltage for a length of 2 μm is 30 V. The amount of impact ionization required to cause snapback depends on N_B which would influence the maximum voltage across the resistor before snapback occurs.

Snapback in the resistor could also occur due to heating in the saturation region [Amerasekera93]. In fact, for more highly doped resistors, self-heating is more likely to result in snapback than avalanche breakdown. Once snapback occurs the current in the resistor is carried by both holes and electrons. Further heating will eventually result in the melt temperature of silicon being reached and damage occurring. Analytical studies and simulations have also shown that current constrictions in the resistor may also occur in the negative resistance region [Khurana66][Hower70][Yang93]. Current constrictions would result in filaments forming and silicon melting taking place at lower injection current levels.

3.3 DIODES

The simplest active voltage clamping device is the diffusion diode. Bias conditions and the associated I–V curve for a forward-biased diode are shown in Figure 3.2 and for a reverse-biased diode in Figure 3.3. In the forward direction the diode begins appreciable current conduction at ≈ 0.5 V, and has an on-resistance R_{on} of between 1 Ω and 5 Ω. Under reverse-biased conditions, the current conduction begins when the junction goes into avalanche breakdown. The avalanche breakdown voltage, BV_{av}, is a function of the n and p doping concentrations and in a submicron process lies between 10 V and 20 V.

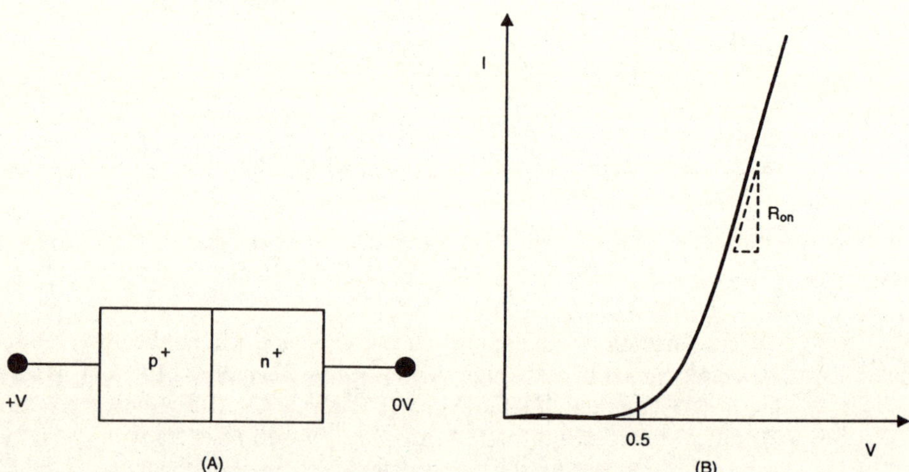

Figure 3.2   (A) Forward-bias condition for a simple *pn* diode. (B) Forward-biased I–V curve for a *pn* diode. Appreciable current conduction is observed at ≈ 0.5 V and the dynamic on-resistance is given by $R_{on} = \Delta V/\Delta I$.

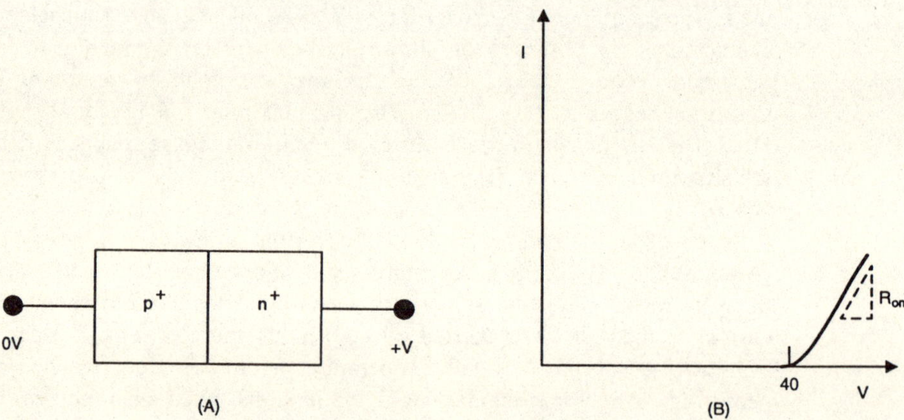

Figure 3.3   (A) Reverse-bias condition for a simple *pn* diode. (B) Reverse-biased I–V curve for a *pn* diode. Avalanche breakdown in this case is shown to begin at 40 V with a dynamic on-resistance of R_{on}.

3.3.1 Forward bias

The ideal diode equation gives the current flowing through the diode as a function of applied bias:

$$I = I_0(T) \cdot \left[\exp\left(\frac{qV}{kT}\right) - 1 \right] \qquad (3.9)$$

$k = 1.38 \times 10^{-23}$ J/K is Boltzmann's constant, V is the voltage across the junction and T is the temperature. I_0 is the saturation current given by

$$I_0(T) = \frac{qAD(T)n_i^2(T)}{N_B L_d} \qquad (3.10)$$

and $D(T)$ and N_B are the minority carrier diffusion coefficient and the doping concentration respectively. L_d is the diffusion length. This equation is applicable to diodes operating in the regime of low level injection. At room temperature the current increases by an order of magnitude for each 60 mV increase in forward bias. At high current levels, the forward diode current equation is modified to become

$$I_{high} = I_0' \exp\left(\frac{qV}{2kT}\right) \qquad (3.11)$$

where

$$I_0' = \left(\frac{2N_B}{n_i}\right) I_0$$

$$= \left(\frac{2qDn_i}{L_d}\right) I_0 \qquad (3.12)$$

which is independent of the background (well) doping concentration N_B. This is an indication that the well region has become conductivity modulated. In a conductivity modulated region, the n and p concentrations are both higher than the background doping concentration. Hence,

$$p = n = n_i \exp\left(\frac{qV}{2kT}\right) \qquad (3.13)$$

and the conductivity is a function of the number of carriers crossing the junction rather than the doping level of the well region. The transition between low and high level injection occurs at

$$V_0 = \frac{2kT}{q} \ln\left(\frac{2N_B}{n_i}\right) \qquad (3.14)$$

For the case of $N_B = 10^{17}/\text{cm}^3$ at room temperature, $V_0 \sim 0.8$ V. This high level injection region is rarely observed because the resistance of the heavily doped regions becomes important at about the same bias.

In lightly doped substrates and diodes built in a well, the current flow is almost entirely lateral between the adjacent highly doped regions as indicated in Figure

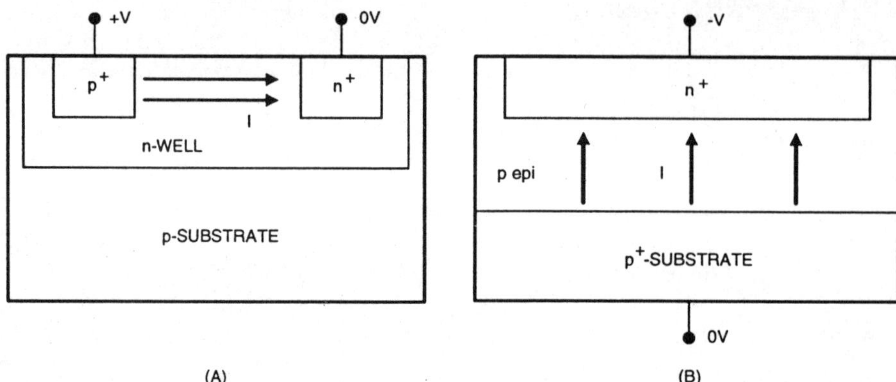

Figure 3.4   (A) Forward-biased current in an *n*-well diode. The current flow is almost entirely lateral. (B) Forward-biased current flow in a substrate diode. The current flow is vertical.

3.4(A). The diode area is defined by the junction sidewall area, i.e. proportional to the junction depth and the peripheral length. Diodes built in thin epitaxial substrates will have current flow paths which are vertical as shown in Figure 3.4(B). In this case the bottom wall area of the junction defines the diode area, i.e. the diode area is proportional to the area of the diffusion.

3.3.2 Reverse bias

When a diode junction is reverse-biased more than a few tenths of a volt, the reverse current is defined by

$$I_R = \frac{qADN_CN_V}{L_dN_B}\exp\left(\frac{-E_g}{kT}\right) + \frac{qW}{\tau_e}\sqrt{N_CN_V}\exp\left(\frac{-E_g}{2kT}\right) \tag{3.15}$$

where N_C and N_V are the density of states in the conduction band and the valence band respectively, W is the width of the depletion region and τ_e is the effective carrier lifetime. The assumption in the above equation is that the field across the junction depletion region at the applied voltage is below the critical field necessary for avalanche breakdown ($\sim 2 \times 10^5$ V/cm). The reverse current is thus solely due to the sum of thermally generated carriers in the depletion region given by the second term on the right-hand side of Equation (3.15) and the diffusion component of the carriers in the neutral region given by the first term on the right-hand side of Equation (3.15). Figure 3.5 shows a plot of $\ln(I_R)$ as a function of $1/kT$. At low temperatures I_R will be dominated by the thermal generation, and shows a slope of (1/2). At higher temperatures the diffusion current dominates and I_R has a slope of 1.

As the reverse voltage is increased and the electric field across the junction becomes of the order of 10^5 V/cm, the carriers in the depletion region can impart enough energy in a collision with the lattice to generate electron-hole pairs which can become free carriers. These new carriers in turn are accelerated, collide with the

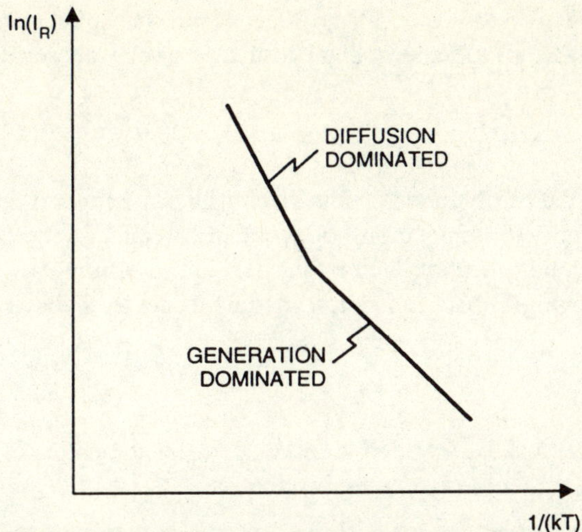

Figure 3.5 Current in a reverse-biased junction, ln(I_R), as a function of temperature, 1/(kT). The diffusion-dominated region where ln(I_R) $\propto$ 1/(kT) is observed at higher temperatures, and the generation-dominated region where ln(I_R) $\propto$ 1/($2kT$) is observed at lower temperatures.

lattice and create more carriers. The process is called avalanche multiplication. The hole and electron currents, I_{p0} and I_{n0}, flowing into a high field region are multiplied so that the currents exiting the region are:

$$I_p = I_{p0} + \alpha_p I_{p0} + \alpha_n I_{n0} \tag{3.16}$$

$$I_n = I_{n0} + \alpha_p I_{p0} + \alpha_n I_{n0} \tag{3.17}$$

$\alpha_{n,p}$ are the electron and hole impact ionization coefficients and are temperature dependent. Okuto and Crowell [Okuto75] give an empirically determined form for the impact ionization coefficients for electrons and holes, $\alpha_{n,p}$ /cm, as a function of temperature:

$$\alpha_{n,p} = A_{n,p} \cdot \{1 + C_{n,p} \cdot 10^{-4}(T - 300)\} \cdot E \cdot \exp\left(\frac{-B_{n,p}^2 \cdot [1 + D_{n,p} \cdot (T - 300)]^2}{E^2}\right) \tag{3.18}$$

$A_n = 0.426$ /V, $A_p = 0.243$ /V, $B_n = 4.81 \times 10^5$ V/cm, $B_p = 6.53 \times 10^5$ V/cm, $C_n = 3.05 \times 10^{-4}$, $C_p = 5.35 \times 10^{-4}$, $D_n = 6.86 \times 10^{-4}$, and $D_p = 5.87 \times 10^{-5}$ are the coefficients for electrons and holes; E is in V/cm. The above equations can be simplified to obtain $\alpha_{n,p}$ by reducing to

$$\alpha_{n,p} = C_1 \cdot \exp\left(\frac{-C_2}{E}\right) \tag{3.19}$$

where $C_2 = E_g/q\lambda$ and λ is the mean free path of the carrier [Crowell66]:

$$\lambda = \lambda_0 \tanh\left(\frac{E_{r0}}{2kT}\right) \tag{3.20}$$

$\lambda_0 \approx 50$ Å and $E_{r0} = 50$ meV are λ and the optical phonon energy, E_r, at 0 K. A reasonable fit to experimental results has been obtained by using [Grant73]

$$\alpha_{n,p} = C_1 \cdot \exp\left(\frac{-b(T)}{E}\right) \tag{3.21}$$

where the coefficient C_1 remains constant as a function of temperature and the major variation with temperature is assumed to occur in the exponent.

The multiplication factor for holes and electrons, $M_{n,p}$, is an important parameter in applying impact ionization models to device behavior.

$$M_{n,p} = \frac{I_{n,p}(out)}{I_{n,p}(in)} \tag{3.22}$$

$I_{n,p}(out)$ and $I_{n,p}(in)$ define the currents at the edges of the depletion region.

For $\alpha_n \approx \alpha_p$, $M_{n,p}$ can be written as

$$M_{n,p} = \frac{1}{1 - \int_0^W \alpha \, dx} \tag{3.23}$$

Avalanche breakdown occurs at the voltage when $M_{n,p}$ approaches infinity, that is when

$$\int_0^W \alpha \, dx = 1 \tag{3.24}$$

A relationship between M and the voltage across the junction has been empirically determined to be [Miller57]

$$M = \frac{1}{1 - (V/BV)^n} \tag{3.25}$$

where V is the voltage across the junction and BV is the avalanche breakdown voltage. The increase in current with applied voltage is very sharp as V approaches BV. The only limiting factor is the resistance of the neutral regions outside the depletion region. The temperature dependence of α is such that as T increases, the impact ionization decreases and M goes down. Hence the avalanche breakdown voltage increases with temperature.

3.3.3 p–i–n Diode

Most diodes built in advanced IC processes have highly doped p and n diffusions separated by a low-doped region of either p type in n to substrate diodes or n type in the case of the n-well diode. At low current levels the low-doped or intrinsic (i) region acts as a resistor in series with either the anode or cathode of the diode. However, at high current levels, the i-region becomes strongly conductivity modulated, i.e. the n and p concentrations are significantly greater than their equilibrium values.

Consider the simple p^+–n–n^+ diode shown in Figure 3.6. As the voltage at the p^+ diffusion, V, is increased, the current increases in accordance with the diode

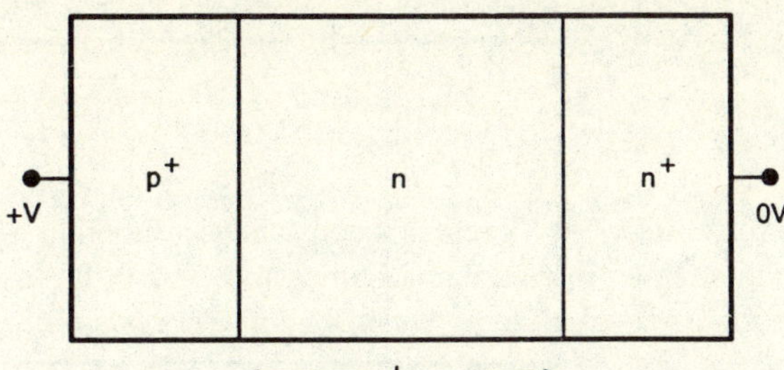

Figure 3.6   A simple p^+–n–n^+ diode in forward bias.

equation. Holes entering the n region recombine with electrons there and the current flow is essentially carried by electrons. The E-field in the n region builds up until the carrier velocities reach v_s and the current saturates. At the same time the field now allows holes to cross the n region and reach the n/n^+ junction, reducing the space charge at this junction and allowing more electrons to be injected into the n region. The voltage required to maintain the same I can begin to decrease and a negative resistance region is observed as the voltage snaps back to a lower sustaining level [Ghandhi77][Chatterjee88]. The on-resistance of the p–i–n diode at high current levels is much lower than at low current levels which is beneficial for ESD protection circuits.

3.4 TRANSISTOR OPERATION

3.4.1 Bipolar Transistors

In advanced MOS and bipolar technologies, protection devices are commonly based on some form of bipolar transistor action. A bipolar transistor in normal operation has a forward-biased junction which enables minority carriers to be injected into the vicinity of a reverse-biased junction. Since currents in reverse-biased junctions are carried by minority carriers, the increase in minority carrier concentration at the junction edge will result in an increase in the current across the junction. The current flow across the reverse-biased junction can be modulated by controlling the injection of minority carriers from the forward-biased junction, and this forms the basis of the bipolar transistor.

A simple npn bipolar structure is shown in Figure 3.7. This is a three-terminal device consisting of two pn junctions. Electron injection takes place from the 'emitter' n region when a positive voltage V_{be} is applied across the junction. The carriers are injected into the 'base' p region. The current in the reverse-biased junction will be enhanced if the width of the base is small enough that there is no significant loss of electrons in the base due to recombination with holes. Electrons cross the reverse-biased junction into the 'collector' n region. The number of carriers reaching the reverse-biased collector junction is determined by the width of the base region, W. The hole current density in the transistor, J_p, is negligible and

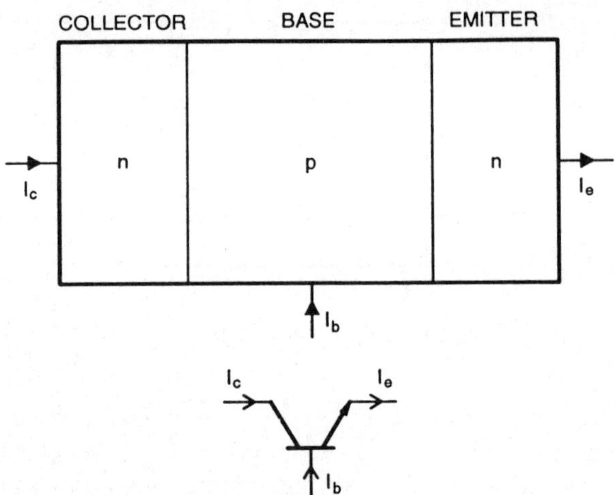

Figure 3.7 A simple *npn* transistor showing the collector current (I_c), emitter current (I_e) and base current (I_b).

the electron current density, J_n, is given by the equation

$$J_n = J_s \left[\exp\left(\frac{qV_{bc}}{kT}\right) - \exp\left(\frac{qV_{be}}{kT}\right) \right] \qquad (3.26)$$

V_{bc} is the base-collector voltage and J_s is the saturation current density given by

$$J_s = \frac{qn_i^2 \tilde{D}}{N_B} \qquad (3.27)$$

n_i is the intrinsic carrier concentration and N_B is the total number of impurities per unit area of the base (*Gummel number*). $\tilde{D}$ is the effective minority carrier diffusion coefficient which assumes that the diffusion coefficient D is not strongly dependent on position.

To turn on an *npn* bipolar transistor, either V_{bc} or V_{be} needs to be positive and greater than kT/q. J_n then becomes a function of the most positive voltage. Under active bias the term $\exp(qV_{bc}/kT)$ is negligible and the collector current in the *npn* bipolar transistor is given by

$$I_c = I_s \exp\left(\frac{qV_{be}}{kT}\right) \qquad (3.28)$$

where $I_s = J_s \times A$ and A is the base-emitter junction area.

Figure 3.8 shows how I_c varies with V_{ce} for an *npn* transistor as the base current I_b is increased. Initially, with $I_b = 0$, the only current flow is that across the reverse-biased collector-base junction until breakdown of the junction takes place at BV_{cbo}. As I_b is increased, i.e. V_{be} is made more positive, the transistor turns on. Higher I_b results in higher I_c in both the saturation and linear regions.

The *current gain* of the transistor is a measure of its effectiveness and is determined by the ratio of the output current to the input current for a specific bias condition. Since I_c is an exponential function of V_{be}, the smaller the current between the base and emitter for a given V_{be} the more effective the transistor. The base-emitter

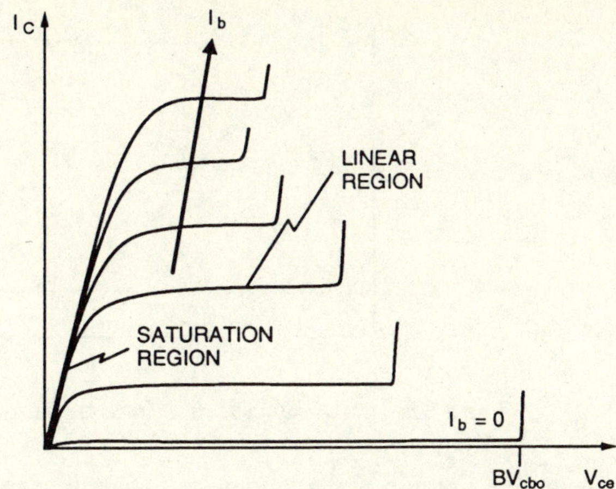

Figure 3.8 I_c as a function of the collector-emitter voltage V_{ce} for an *npn* transistor with increasing I_b. BV_{cbo} indicates the collector-base breakdown voltage with the emitter open-circuit.

current, I_b, consists of (1) the current due to recombination of injected electrons with holes in the base, (2) the current due to recombination in the space-charge region, and (3) the current due to hole injection from the base into the emitter. The sum of these currents defines I_b. The gain is improved if recombination is reduced (i.e. lifetime increased) in the base for electrons and in the emitter for holes.

The two important current gain definitions for circuits used under ESD type conditions are the common-base current gain, α, and the common-emitter current gain, β. α and β are given by

$$\alpha = \frac{\partial I_c}{\partial I_e} \sim \frac{I_c}{I_e} \tag{3.29}$$

$$\beta = \frac{\partial I_c}{\partial I_b} \sim \frac{I_c}{I_b} \tag{3.30}$$

α and β are related by

$$\beta = \frac{\alpha}{1 - \alpha} \tag{3.31}$$

Typically, $\alpha \approx 1$ and $\beta \gg 1$.

Two other parameters of importance are the base transport factor α_T and the emitter efficiency γ which is the effectiveness of the emitter junction in injecting electrons into the base. α_T is a measure of the loss of carriers due to recombination in the base region:

$$\alpha_T = \frac{I_n(W)}{I_n(0)} \tag{3.32}$$

$$\approx 1 - \frac{W^2}{2L_B^2} \tag{3.33}$$

L_B is the minority carrier diffusion length in the base.

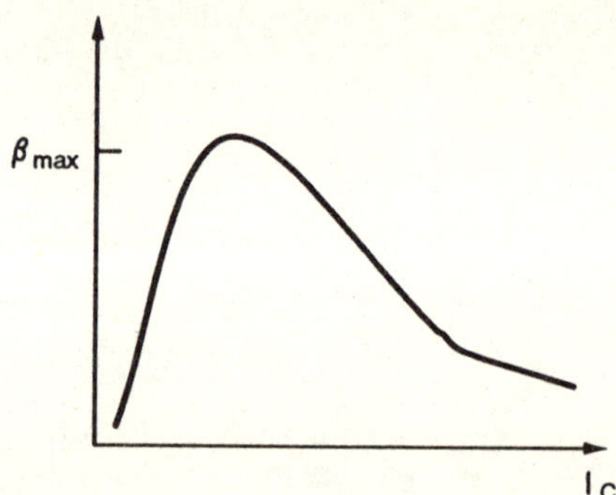

Figure 3.9 β as a function of I_c for a bipolar transistor.

γ is given by the ratio of the incremental electron current from the emitter to the incremental total electron current,

$$\gamma = \frac{I_{ne}}{I_{ne} + I_{pe}} \tag{3.34}$$

I_{ne} is the electron current across the emitter-base junction and I_{pe} is the hole current across the emitter-base junction.

The recombination current only flows between the emitter and the base. I_c consists almost entirely of the collected electrons which were injected at the base-emitter junction. As V_{be} is decreased, I_c follows Equation 3.28 until injection is so low that generation in the space-charge region begins to dominate. At low currents, therefore, I_c is a smaller fraction of I_e and β is low. The variation of β with collector current is shown in Figure 3.9. In the ideal region, β is more or less constant, and then as I_c increases further, β begins to decrease rapidly. β degradation at high I_c occurs when the injected minority carrier density into the base approaches the majority carrier density thereby increasing the majority carrier charge in the base. Hence, the injected carriers effectively increase the base doping, which reduces the emitter efficiency and β. A second effect at high injection levels is current crowding at the emitter which is essentially due to two-dimensional effects. The flow of majority carriers in the base region leads to a potential drop across the width of the emitter and causes variation in the forward-biased voltage across different regions of the emitter. This results in a variation in the injected current density across the emitter junction. For example, a lateral ohmic drop of ≈ 26 mV will result in a reduction in the emitter current density of $1/e$. At high currents, therefore, the effective emitter area is reduced and this contributes to the reduction in β.

An important parameter for time-dependent operation of bipolar transistors is the *base transit time*, τ_B [Muller86][Krieger89]. τ_B determines how long it takes for electrons injected at the emitter to reach the collector junction and turn on the transistor. If the injected minority carriers into the base have a charge Q_{nB}, then

$$Q_{nB} = \int_0^W qAn'(x)\, \mathrm{d}x \tag{3.35}$$

where A is the area of the emitter and $n'(x)$ is the concentration of the excess electrons in the base at a point x. For a collector current, I_C, τ_B is given by

$$\tau_B = \frac{Q_{nB}}{I_C} \tag{3.36}$$

If $n'(x)$ is assumed to be linear across W and substituting for I_C then

$$\tau_B = \frac{W^2}{2\tilde{D}_n} \tag{3.37}$$

Under high current levels, a field is set up between the emitter and collector which aids the flow of minority carriers towards the collector and [Muller86]

$$\tau_B = \frac{W^2}{4\tilde{D}_n} \tag{3.38}$$

Typical values of τ_B for a 1 μm wide base are about 250 ps which are well within the rise times associated with ESD stress currents.

3.4.2 MOS transistors

In advanced CMOS circuits, nMOS and pMOS transistors in the input and output buffers will be subjected to the high currents and voltages associated with an ESD stress. A schematic cross-section of an nMOS transistor is shown in Figure 3.10 and consists of two n diffusion regions in a p substrate. A pMOS transistor is the complement of the nMOS and the description presented here is applicable to the pMOS with changes in the polarity of the majority and minority carriers. Upon application of a positive voltage, V_D, at the drain (D), with the gate (G), source (S) and substrate (B), connected to zero volts, no current will flow until the reverse-biased drain-substrate junction goes into avalanche breakdown as depicted by the curve $V_G = 0$ V in Figure 3.11. When a positive voltage, V_G, is applied at the gate, the p region between the drain and source first becomes depleted and then, as V_G is increased further, the p-region becomes inverted and an n channel is formed. The transistor is now in the on-state. The gate voltage at which the transistor turns on is called the threshold voltage V_T. The relationship between V_D and the drain current I_D for different V_G is shown in Figure 3.11. In the linear region indicated in the Figure, the drain current is given by

$$I_D = \frac{aW}{L} \cdot (V_G - V_T) \cdot V_D - b \cdot V_D^2 \tag{3.39}$$

a is a constant dependent on the electron mobility in the channel and the capacitance between the gate and the semiconductor. W is the width of the transistor, L is the channel length and b is a constant dependent on the gate oxide capacitance and channel doping concentration.

In the saturation region, the current is given by

$$I_{Dsat} \approx \frac{b'W}{L}(V_G - V_T)^2 \tag{3.40}$$

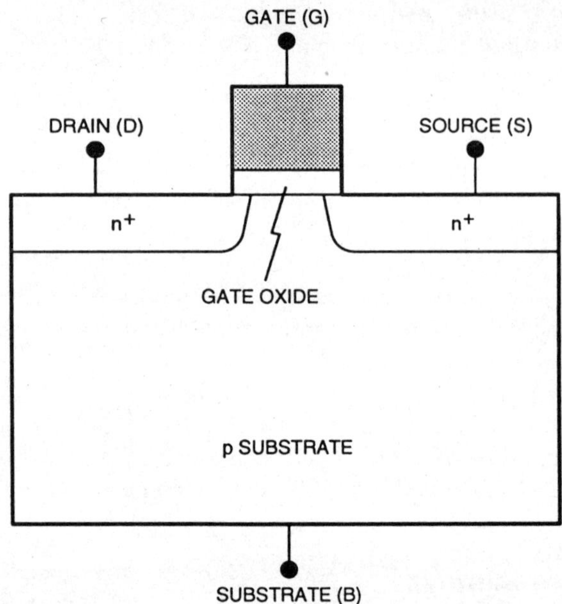

Figure 3.10 Schematic cross-section of an nMOS transistor.

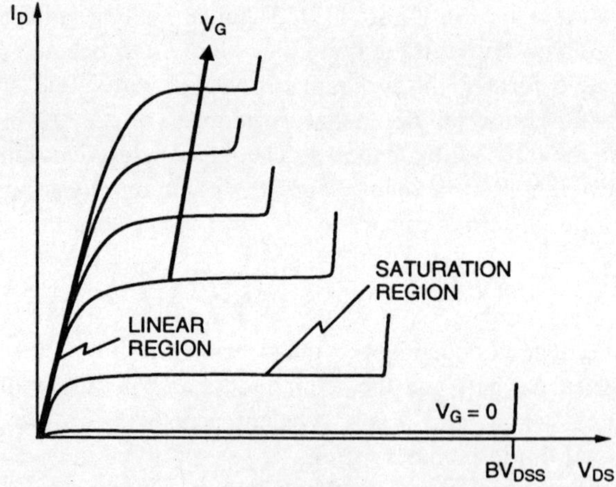

Figure 3.11 The drain current, I_D, of an nMOS transistor as a function of the drain-source voltage, V_{DS}, with increasing gate voltage, V_G. BV_{DSS} is the drain-substrate breakdown voltage with $V_G = 0$.

b' is similar to b but includes a doping dependent parameter [Sze81]. From these two equations we see that the linear current is a linear function of the gate voltage, while the saturation current is a quadratic function of V_G.

3.4.3 Avalanche conditions

Both bipolar and MOS transistors in ICs are eventually biased into avalanche during high current injection. In this section we will briefly outline the main features in a transistor operating with one avalanching junction. The analysis is based on the work of Dutton [Dutton75] and Reisch [Reisch92]. For simplicity consider the 1-D bipolar transistor shown in Figure 3.12. α is the common-base gain and M is the multiplication factor in the collector-base depletion region. The collector current, I_C, is given by

$$I_C = \alpha I_E + (M - 1)\alpha I_E + M I_{CO}$$
$$= \alpha M I_E + M I_{CO} \tag{3.41}$$

where I_E is the emitter current and I_{CO} is the thermal generation current across the depletion region as described for reverse-biased junctions in Equation (3.15). Since the electrons recombining in the base must equal the holes recombining in the base,

$$(1 - \alpha)I_E = M I_{CO} + (M - 1)\alpha I_E + I_B \tag{3.42}$$

and I_B is the base current being injected into the transistor. Now since

$$(1 - \alpha) = \frac{\alpha}{\beta} \tag{3.43}$$

and

$$\alpha I_E = \frac{I_C}{M} - I_{CO} \tag{3.44}$$

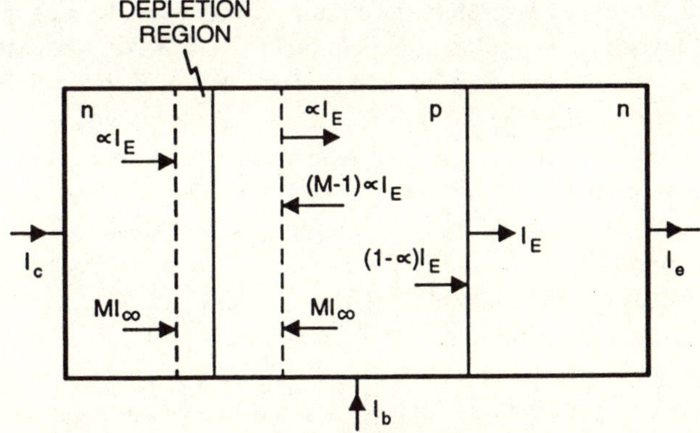

Figure 3.12 Simple 1-D model of a bipolar transistor with avalanching in the reverse-biased collector-base depletion region. M is the avalanche multiplication factor in the depletion region and α is the common-base current gain.

we get

$$I_C = \frac{M\beta I_B + MI_{CO}(1+\beta)}{1 - \beta(M-1)} \tag{3.45}$$

As the avalanche generation at the collector-base junction, increases the hole current entering the base terminal $I_B \rightarrow 0$ and eventually becomes negative. The condition for $I_B \leq 0$ and $I_C > 0$ is given by

$$\beta(M-1) \geq 1 \tag{3.46}$$

$$\beta M \geq (\beta + 1) \tag{3.47}$$

In the open-base condition $I_B = 0$ and Equation (3.45) reduces to

$$I_C = I_E = I = \frac{MI_{CO}}{1 - \alpha M} \tag{3.48}$$

as given by Sze [Sze81]. Equation (3.48) describes the current in a floating base transistor and combined with Equation (3.25) can be used to determine the collector-emitter breakdown voltage, BV_{ceo}, in terms of the collector-base breakdown voltage, BV_{cbo}:

$$BV_{ceo} = BV_{cbo}(1-\alpha)^{1/n} \tag{3.49}$$

3.5 TRANSISTOR OPERATION UNDER ESD CONDITIONS

3.5.1 Bipolar transistors

The bipolar transistor can be designed to carry high currents under normal conditions, but such devices require a high collector-emitter voltage V_{ce} and dissipate a lot of power. ESD protection circuits using such devices are used in high-voltage and high-power applications [Corsi93]. However, transistors designed for low voltage (5 V) applications cannot carry large currents and the behavior deviates from the standard operation.

Under ESD conditions, the base terminal is connected to the emitter terminal either directly or through an external resistor R_{ext} as shown in Figure 3.13. Two turn-on modes are possible; the first bias condition is shown in Figure 3.13 and the second bias condition is shown in Figure 3.14. The first bias type operates through self-biasing using the internal avalanche generation of carriers to turn on the bipolar transistor. As will be discussed below, the bipolar is triggered at the collector-base breakdown voltage which is $\approx BV_{cbo}$. In the second bias type the transistor is provided some forward biasing by means of an external current source, T, in Figure 3.14, and the voltage drop across the resistor R_{ext}. The advantage of this schematic over the first is that it reduces the voltage at which the bipolar turns on in the self-biasing mode [Chatterjee91A][Amerasekera92]. We will discuss the mechanisms of the bipolar operation in more detail in this section.

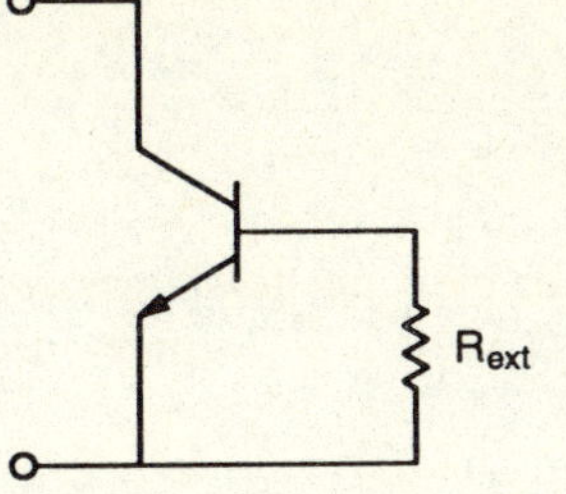

Figure 3.13 Schematic of an *npn* transistor operating in the self-biasing mode. R_{ext} is an external resistor connected between the base and the emitter to ensure *npn* turn-on when the collector voltage is increased.

In the first type, a positive current is forced into the collector and Figure 3.15 shows the variation of the forced I_c with the collector-emitter voltage, V_{ce}. Initially the reverse-biased collector-base junction is a high impedance and the reverse current is given by Equation (3.15). The collector voltage rises until it reaches the collector-base junction breakdown voltage, BV_{cbo}. Impact ionization now takes place in the junction and electron-hole pairs are generated. The electric field is such that the electrons enter the collector and increase I_c, and the holes drift to the base terminal generating a negative base current. The hole current into the base contact will result in a potential drop in the base, and when the emitter-base voltage reaches $\sim$0.5 V the *npn* begins to turn on. (*Note*: The turn-on voltage is actually not a precise value. Under ESD conditions, turn-on will occur when V_{be} is sufficiently forward-biased to support the ESD stress current injected at the collector. We will use $V_{be} \approx 0.5$ V in order to simplify the analysis.) The generated hole current, I_B, required to forward-bias the emitter-base junction will depend on the internal base resistance, R_B, as well as R_{ext}. I_B required to turn on the *npn* can be calculated from

$$V_{be} = I_B \cdot (R_B + R_{ext}) \qquad (3.50)$$

$$V_{be} > 0.5 \text{ V} \qquad (3.51)$$

The function of the generated I_B under ESD conditions is to increase V_{be} and forward-bias the emitter-base junction thus providing the electrons necessary for the *npn* action to take place. Once the *npn* turns on, the electron current reaching the

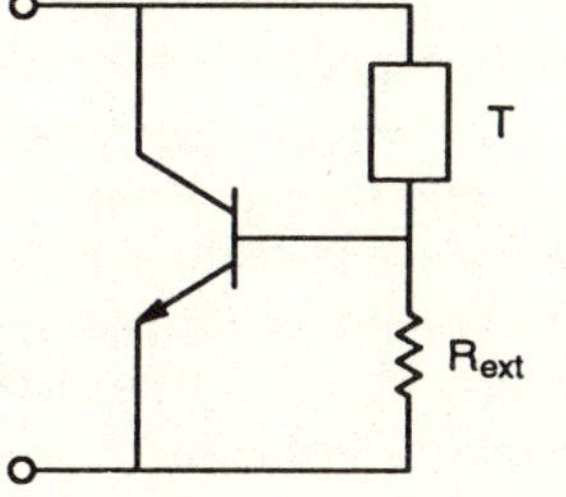

Figure 3.14 Schematic of an *npn* connected in the triggered bias condition. R_{ext} is an external resistor and the *npn* is triggered when the current through the trigger element, T, is sufficient to adequately forward-bias the emitter-base junction of the *npn*.

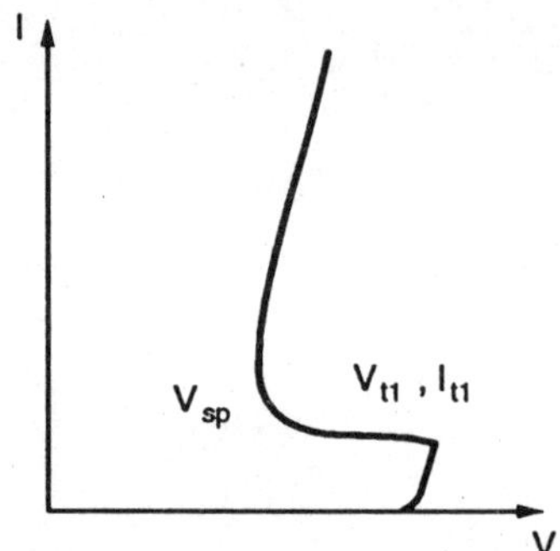

Figure 3.15   High current *I–V* curve for an *npn* transistor.

reverse-biased collector-base junction increases the number of generated electron-hole pairs, since [Dutton75][Reisch92]

$$I_B \propto (M - 1) \cdot I_e \tag{3.52}$$

M is a function of the voltage across the junction, and as I_e increases M can decrease and V_{cb} can also decrease. The voltage at which V_C begins to decrease is the snapback trigger voltage denoted by V_{t1} in Figure 3.15. The snapback trigger current is given by $I_c = I_{t1}$. The multiplication factor, M, can be estimated from

$$\frac{I_B}{I_{t1}} = 1 - \frac{1}{M} \tag{3.53}$$

V_{ce} now decreases until eventually a stable condition is reached whereby the generated I_B satisfies Equation (3.50). The condition for snapback which requires that I_B flows out of the base terminal and I_C is positive has been previously defined in Equation (3.46) as

$$\beta(M - 1) \geq 1 \tag{3.54}$$

V_{be} necessary to support the injected I_c is obtained from Equation (3.28). Further increase in I_c results in an increase in V_{ce} as conductivity modulation of the internal base resistance necessitates an increase in I_B to sustain the transistor in the on-condition. The snapback holding voltage is given by V_{sp}, and is a function of the base-width, W, the multiplication factor, M, for the collector-base junction, R_B and R_{ext}.

In the second operating mode (Figure 3.14), T is a current source formed by either a transistor or a diode operating in reverse-bias with a breakdown voltage much less than BV_{cbo}. The current from T goes through the resistor R_{ext} and increases V_{be}. As V_{be} becomes more positive, the electrons entering the base from the emitter can contribute to the avalanche generation at the collector junction. Thus the voltage (i.e. M) required for a given I_B is lower and, since V_{be} is already positive, the I_B required to fully turn on the bipolar is reduced. V_{t1} is, therefore, lower than for the first bias condition without external triggering. The lower V_{t1} is extremely desirable in ESD protection circuits because it ensures that the protection device will trigger before the device being protected.

3.5.2 MOS transistors

MOS transistors designed for ICs in advanced CMOS processes do not have the capability of carrying amps of current. In addition, the thin gate oxides of the order of 100 Å mean that high voltages cannot be sustained in these technologies. Under ESD stress conditions the MOS transistor behavior changes drastically from normal operation. The actual conduction mechanism is that of bipolar action in the parasitic lateral bipolar shown in Figure 3.16 (see for example [Krieger89]).

As the drain voltage is increased from 0 V, the reverse-biased drain-substrate junction is in high impedance and the only current is the reverse current given by Equation (3.15). Eventually the E-field across the depletion region in the drain substrate junction becomes high enough to begin avalanche multiplication and the junction goes into avalanche breakdown with the generation of electron-hole pairs. The generated electrons are swept across the drain towards the drain contact, adding to the drain current. The generated holes drift towards the substrate contact giving rise to a substrate current, I_{sub}, similar to the case of the base current, I_B, for the bipolar transistor. The effective substrate resistance is denoted by R_{sub}. Figure 3.17 shows schematically the paths of the holes and electrons in the nMOS. As I_{sub} increases, the potential at the source-substrate junction increases thereby

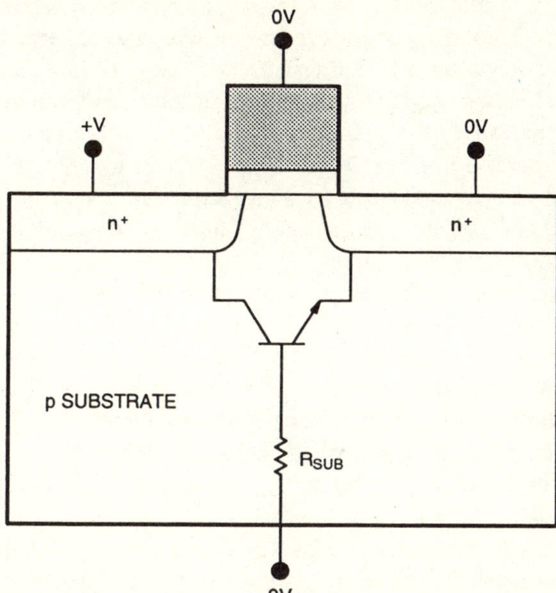

Figure 3.16 Cross-section of nMOS transistor showing a schematic of the lateral *npn* transistor. R_{sub} is the effective substrate resistance.

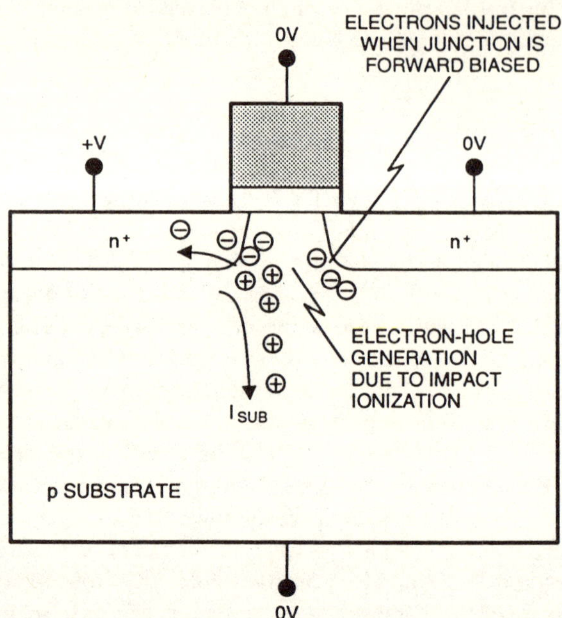

Figure 3.17 Cross-section of nMOS transistor showing the paths of the holes and electrons generated by impact ionization. The hole current results in I_{sub}.

forward-biasing this junction and electrons are emitted into the substrate. When the electron current density from the source begins to contribute to the drain current the parasitic bipolar transistor can be considered to be turned on. Figure 3.18 shows the $I–V$ curve of an nMOS transistor with gate, source and substrate at zero volts. A positive voltage or current is applied at the drain terminal. V_{t1} is the turn-on voltage of the parasitic bipolar transistor and the associated current is I_{t1}. The drain of the nMOS becomes the collector of the bipolar, the source of the nMOS becomes the emitter of the bipolar and the substrate is the base. The turn-on time of the parasitic bipolar is defined by the base transit time τ_B. For a 1 μm channel length, $\tau_B \approx 250$ ps [Krieger89].

Once the lateral bipolar transistor turns on, the operating mechanism is similar to that of the *npn* bipolar transistor described in the previous section. The drain voltage decreases and a negative resistance region is observed due to the availability of more carriers for multiplication until a minimum voltage, V_{sp}, is reached. The $I–V$ curve now shows a positive resistance as further increase in the injected current results in conductivity modulation of the substrate (base) region which reduces the intrinsic substrate resistance. A higher I_{sub} is required to maintain the transistor in the on-condition.

As in the case of the bipolar transistor, when the MOS transistor is turned on, the transistor current reduces the voltage at which the drain–substrate breakdown occurs. This feature has been used in the design of ESD protection circuits using nMOS transistors, as will be discussed in a later chapter.

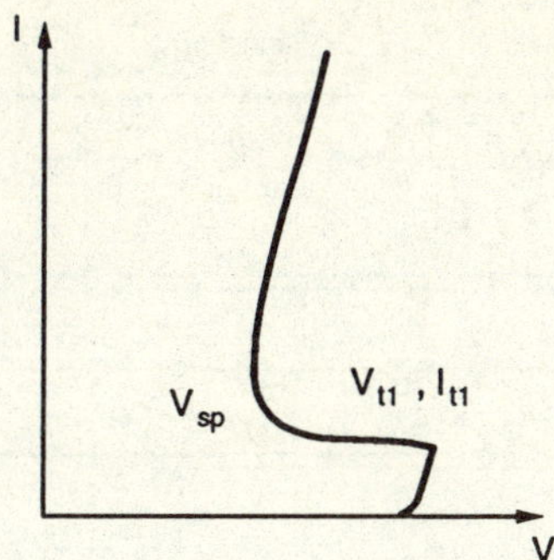

Figure 3.18   High current *I–V* curve of an nMOS transistor with voltage applied at the drain and with the gate, source and substrate at zero volts.

In general the high current operation of the parasitic bipolar transistor in a MOS device is exactly the same as that of the 'vertical' bipolar transistor described in the previous section. There are some features of the parasitic bipolar transistor which are different and need to be emphasized because they affect the high current performance [Lindmayer67]. These are related to the fact that the transistor is a lateral device rather than a vertical device. In a lateral bipolar transistor, the majority of the electrons reaching the collector junction are emitted from the emitter junction sidewall which results in a very small 'intrinsic' base area [Verdonckt91]. The high current is confined to a very small region of the emitter and base regions which will lead to large power density in these regions and hence higher temperatures. I_{sub} initially needs to forward-bias a small region of the source–substrate junction to turn on the bipolar. However, for better ESD performance a larger emitter area is preferred. In short-channel MOS transistors there is some influence of the drain depletion region on the source barrier (also called Drain Induced Barrier Lowering or DIBL) [Sze81]. DIBL will help the bipolar transistor by increasing the effective emitter area [Amerasekera94A][Amerasekera94B]. This will be particularly effective if the source barrier lowering occurs deeper in the junction, allowing the power dissipation to take place deeper in the device and reducing the temperature rise in the device.

3.6 SCR OPERATION

Silicon Controlled Rectifiers (SCRs), also known as thyristors, are devices which are used extensively in power device applications because of the capability to switch from a very high impedance state to a very low impedance state. For the same reason a properly designed SCR can also be a very efficient ESD protection circuit [Avery83][Chatterjee91B]. A cross-section of a simple lateral SCR is shown in Figure 3.19 and essentially consists of a *pnpn* structure. The p^+ diffusion in the *n*-well forms the *anode* of the SCR where holes are injected into the *n*-well. The n^+ diffusion in the *p*-well forms the *cathode* of the SCR from which electrons are

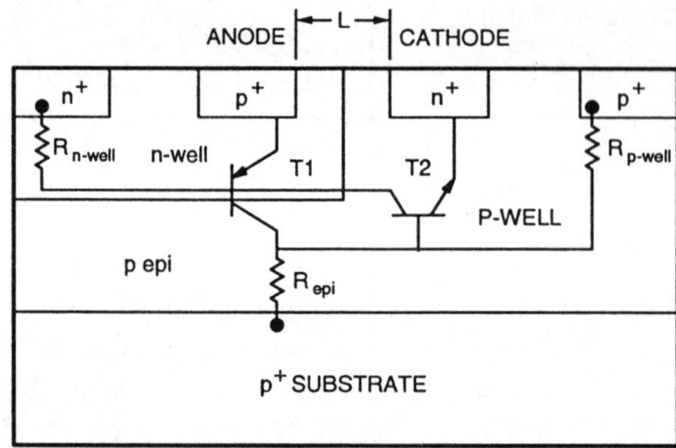

Figure 3.19   Cross-section of a lateral SCR in a CMOS process showing the parasitic *pnp* and *npn* transistors. $R_{\text{n-well}}$ is the *n*-well resistor, $R_{\text{p-well}}$ is the *p*-well resistor, and R_{epi} is the resistance of the epi-layer. The low resistance (5 mΩ-cm) p^+ substrate is assumed to be at 0 V.

injected into the *p*-well. The connection to the *n*-well is made through an n^+ contact in the well, while the p^+ contact in the well is the connection to the *p*-well.

The SCR may be considered as two bipolar transistors. A *pnp* transistor, T1, is formed by the *anode* as emitter, the *n*-well as base and the *p*-well as collector. An *npn* transistor, T2, is formed by the *cathode* as emitter, the *p*-well as the base and the *n*-well as the collector. The SCR may be biased as follows. The *n*-well is connected to a fixed voltage, V_C, the *p*-well and the *cathode* are connected to ground and a voltage V is applied to the *anode*. The I–V curve for the SCR is shown in Figure 3.20. As V goes above V_C the emitter-base junction of the *pnp* is forward-biased and the *pnp* turns on. The current through the *pnp* flows into the *p*-well and forward-biases the emitter-base junction of the *npn*, turning it on. The *npn* current from the *n*-well to the *cathode* now supplies the forward-bias for the *pnp*, the

voltage at the *anode* no longer needs to provide the bias for the *pnp*, and V begins to decrease resulting in a negative resistance region. The minimum value of V is known as the *holding voltage* or V_h and is defined by the amount of current that the

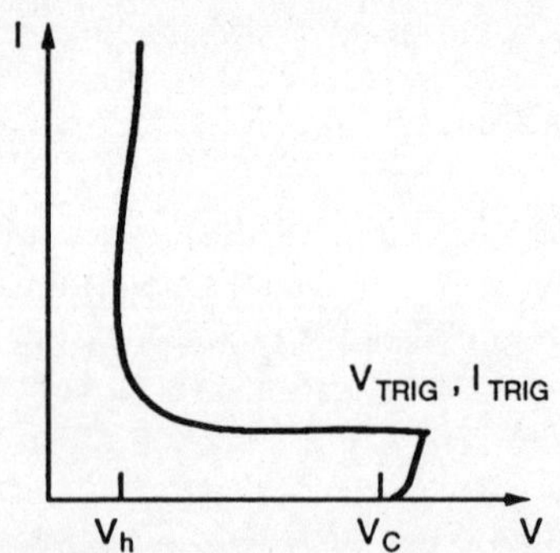

Figure 3.20   High current I–V curve for an SCR.

pnp needs to supply to forward-bias the *npn* and the base-widths of the lateral *npn* and the lateral *pnp*, which is the anode to cathode spacing L in Figure 3.19.

The SCR can be approximated by the equivalent circuit shown in Figure 3.21. T1 and T2 are the *pnp* and the *npn* respectively, and R_{n-well} and R_{p-well} denote the well resistances. When the SCR is in the latched mode the requirement that it stays latched is given by the equation [Estreich81]

$$\beta_{npn} \cdot \beta_{pnp} \geq 1 \tag{3.55}$$

β_{npn} and β_{pnp} are the current gains of the *npn* and *pnp* respectively. It must be noted that although T1 and T2 are shown as discrete transistors, the collector of one is the base of the other. Therefore, it is not correct to use the discrete β's when computing Equation (3.55) above.

The two important parameters for the SCR are the trigger current, I_{trig}, and V_h. I_{trig} is determined by R_{p-well} which in turn is determined by the thickness of the epitaxial layer (when present) and the doping of the *p*-well. V_h is strongly dependent on L, as well as R_{n-well} and is typically between 2 V and 5 V in advanced CMOS processes. Once on, the SCR can be modeled as a *p–i–n* diode [Herlet66][Chatterjee88][Seitchik87]. The region between the anode and cathode is now fully conductivity modulated, and the on-resistance of the SCR is about 1 Ω making it a low power dissipating device ideal for ESD protection circuits.

When used as an ESD protection circuit, the SCR is connected as a two-terminal device, with the anode and *n*-well tied together and the cathode and *p*-well tied together. Triggering now requires avalanche breakdown of the *n*-well to *p* junction. The SCR turns on either when the cathode is forward-biased by the hole current in the *p* region in a manner similar to the triggering of the *npn* in nMOS transistors, or when the *pnp* is turned on by the electron current in the *n*-well. Typically the *npn* gain is an order of magnitude higher than that of the *pnp* at low current levels, and turning on the *npn* is easier to achieve than turning on the *pnp*. The trigger voltage is defined by the avalanche breakdown voltage of the *n*-well to substrate, and the trigger current is the same as for the circuit described above.

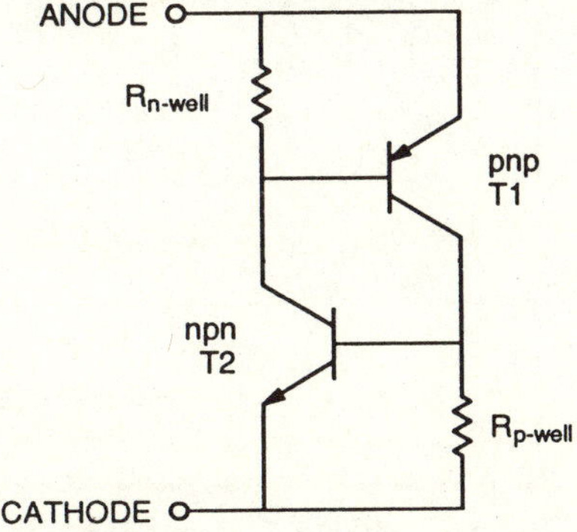

Figure 3.21   Equivalent circuit schematic of the SCR in Figure 3.19.

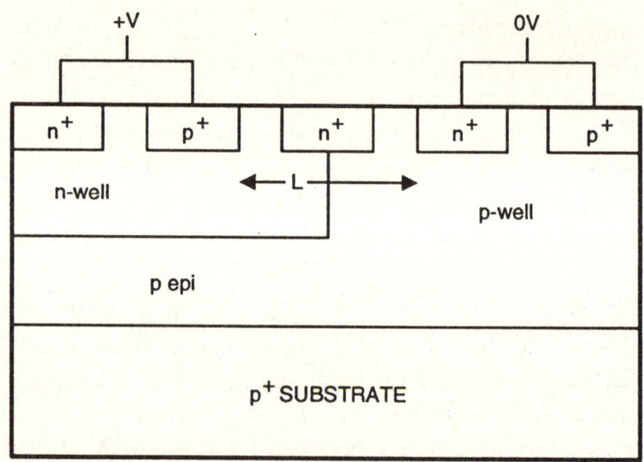

Figure 3.22   Lateral CMOS SCR showing the addition of an n^+ diffusion at the *n*-well edge to reduce the SCR trigger voltage.

The avalanche breakdown voltage of the *n*-well to substrate is about 40 V in an advanced CMOS process. In order to make the SCR a good ESD protection circuit the trigger voltage must be reduced. This is accomplished by using an additional n^+ diffusion at the *n*-well edge as shown in Figure 3.22. The breakdown voltage is now reduced to that of the n^+ to substrate which is about 20 V in a submicron CMOS process. A further reduction of the trigger voltage is achieved by using a gated-diode at the *n*-well edge as shown in Figure 3.23 [Chatterjee91B]. The gate of the nMOS transistor which forms the gated diode is connected to the cathode. The trigger voltages for these devices are between 10 V and 15 V in submicron processes.

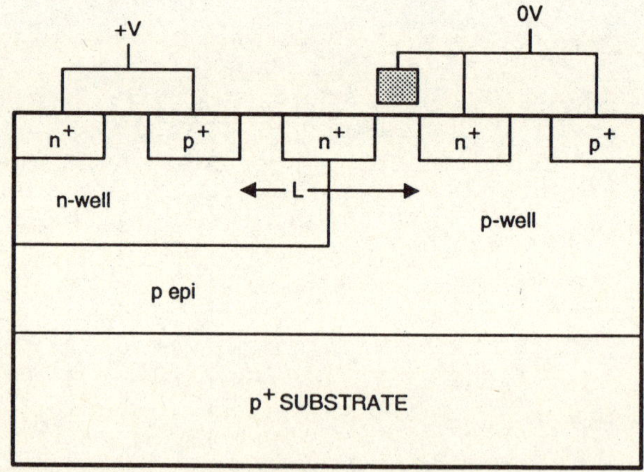

Figure 3.23   Lateral CMOS SCR showing the addition of a gated-diode (nMOS) at the *n*-well edge to reduce the SCR trigger voltage [Chatterjee91B].

REFERENCES

[Amerasekera92]   A. Amerasekera, A. Chatterjee, 'An Investigation of BiCMOS ESD Protection Circuit Elements and Applications in Submicron Technologies', in *Proc. 14th EOS/ESD Symposium*, p. 265–276, 1992.

[Amerasekera93]   A. Amerasekera, M.-C. Chang, J. Seitchik, A. Chatterjee, K. Mayaram, J.-H. Chern, 'Self-Heating Effects in Basic Semiconductor Structures', *IEEE Elec. Dev.*, ED-40, p. 1836–1844, 1993.

[Amerasekera94A]   A. Amerasekera, R. Chapman, 'Technology Design for High Current and ESD Robustness in a Deep Submicron Process', *Elec. Dev. Lett.*, 15, p. 383–385, 1994.

[Amerasekera94B]   A. Amerasekera, J. Seitchik, 'Electrothermal Behavior of Deep Submicron nMOS Transistors under High Current Snapback (ESD/EOS) Conditions', in *Tech. Dig. IEDM*, p. 446–449, 1994.

[Avery83]   L. R. Avery, 'Using SCR's as Transient Protection Structures in Integrated Circuits', in *Proc. 5th EOS/ESD Symposium*, p. 177–180, 1983.

[Caruso74]   A. Caruso, P. Spirito, G. Vitale, 'Negative Resistance Induced by Avalanche Injection in Bulk Semiconductors', *IEEE Trans. Elec. Dev.*, ED-21, p. 578–586, 1974.

[Chatterjee88]   A. Chatterjee, J. A. Seitchik, J.-H. Chern, P. Wang, C.-C. Wei, 'Direct Evidence Supporting the Premises of a Two-Dimensional Diode Model for the Parasitic Thyristor in CMOS Circuits Built on Thin Epi', *IEEE Elec. Dev. Lett.*, EDL-9, p. 509–511, 1988.

[Chatterjee91A]   A. Chatterjee, T. Polgreen, A. Amerasekera, 'Design and Simulation of a 4 kV ESD Protection Circuit', in *Tech. Dig. IEDM*, p. 913–916, 1991.

[Chatterjee91B]   A. Chatterjee, T. Polgreen, 'A Low Voltage Triggering SCR for On-Chip ESD Protection at Output and Input Pads', *IEEE Elec. Dev. Lett.*, EDL-12, p. 21–22, 1991.

[Corsi93]   M. Corsi, R. Nimmo, F. Fattori, 'ESD Protection of BiCMOS Integrated Circuits Which Need to Operate in the Harsh Environments of Automotive or Industrial', in *Proc. 15th EOS/ESD Symposium*, p. 209–214, 1993.

[Crowell66]   C. R. Crowell, S. S. Sze, 'Temperature Dependence of Avalanche Multiplication in Semiconductors', *App. Phys. Lett.*, 9, p. 242–243, 1966.

[Dutton75]   R. W. Dutton, 'Bipolar Transistor Modeling of Avalanche Generation for Computer Circuit Simulation', *IEEE Trans. Elec. Dev.*, ED-22, p. 334–338, 1975.

[Estreich81]   D. B. Estreich, *The Physics and Modeling of Latch-Up in CMOS Integrated Circuits*, PhD Thesis, Stanford University, 1981.

[Ghandhi77]   S. K. Ghandhi, *Semiconductor Power Devices*, New York: Wiley, 1977.

[Grant73]   W. N. Grant, 'Electron and Hole Ionization Rates in Epitaxial Silicon at High Electric Fields', *Sol. St. Elec.*, 16, p. 1189–1203, 1973.

[Herlet66]   A. Herlet, K. Raithel, 'Forward Characteristics of Thyristors in the Fired State', *Sol. St. Elec.*, 9, p. 1089–1105, 1966.

[Hower70]   P. L. Hower, V. G. K. Reddi, 'Avalanche Injection and Second Breakdown in Transistors', *IEEE Trans. Elec. Dev.*, ED-17, p. 320–335, 1970.

[Khurana66]   B. S. Khurana, T. Sugano, H. Yanai, 'Thermal Breakdown in Silicon *p–n* Junction Devices', *IEEE Trans. Elec. Dev.*, ED-13, p. 763–770, 1966.

[Krieger89]   G. Krieger, 'The Dynamics of Electrostatic Discharge Prior to Bipolar Action Related Snapback', in *Proc. 11th EOS/ESD Symposium*, p. 136–144, 1989.

[Lindmayer67]   J. Lindmayer, W. Schneider, 'Theory of Lateral Transistors', *Sol. St. Elec.*, 10, p. 225–234, 1967.

[Miller57]   S. L. Miller, 'Ionization Rates for Holes and Electrons in Silicon;, *Phys. Rev.*, 105, p. 1246–1249, 1957.

[Muller86]   R. S. Muller, T. I. Kamins, *Device Electronics for Integrated Circuits*, 2nd edn, New York: Wiley, 1986.

[Okuto75]   Y. Okuto, C. R. Crowell, 'Threshold Energy Effect on Avalanche Breakdown Voltage in Semiconductor Junctions', *Sol. St. Elec.*, 18, p. 161–168, 1975.

[Reisch92]   M. Reisch, 'On Bistable Behavior and Open-Base Breakdown of Bipolar Transistors in the Avalanche Regime–Modeling and Applications, *IEEE Trans. Elec. Dev.*, ED-39, p. 1398, 1992.

[Seitchik87]   J. Seitchik, A. Chatterjee, P. Yang, 'An Analytical Model of Holding Voltage for Latch-Up in Epitaxial CMOS', *IEEE Elec. Dev. Lett.*, EDL-8, p. 157–159, 1987.

[Sze81]   S. M. Sze, *Physics of Semiconductor Devices*, 2nd edn, New York: Wiley, 1981.

[Verdonckt91]   S. Verdonckt-Vandebroek, S. S. Wong, J. Woo, P. K. Ko, 'High Gain Lateral Bipolar Action in a MOSFET Structure', *IEEE Trans. Elec. Dev.*, ED-38, p. 2487–2496, 1991.

[Yang93]   P. Yang, J.-H. Chern, 'Design for Reliability: The Major Challenge for VLSI', *Proc. IEEE*, 81, p. 730–744, 1993.

4

DESIGN AND LAYOUT REQUIREMENTS

4.1 INTRODUCTION

The requirement for ESD protection circuits to carry currents way beyond the levels for which the elements were initially designed results in regions of high thermal dissipation and high electric fields. A good protection circuit needs to be able to withstand the heating effects, sink the large currents during the ESD event and not be damaged by the ensuing high electric fields. The capability to meet these requirements is critically dependent on the specific design and layout of the protection circuits and the individual elements as well as the circuits which are being protected [Hullet80][Keller81]. Hence, ESD is considered to be very layout intensive. Furthermore, while the design concepts have to change with the technology, the layout techniques must be revised to make them compatible. For example, the same output buffer device will require one type of layout for technologies with no salicided diffusions and a different layout for technologies with salicided diffusions. The layout can also vary for grounded substrate logic chip technologies versus floating substrate DRAM chips technologies. For instance, in DRAMs the proximity effects due to the interaction from neighboring diffusions can lead to failure and force a more careful implementation of the layout [LeBlanc91].

In this chapter we present and discuss the design and layout techniques for input/output pins. The focus will be on circuits designed in advanced CMOS processes but we will also present some typical approaches for bipolar and BiCMOS circuits. Specific examples will be given to illustrate the *right* and *wrong* approaches to protection design and layout.

4.2 DESIGN CONCEPTS

The first requirement for a good protection circuit design methodology is the choice of the appropriate type of protection device that is compatible and/or suitable for the

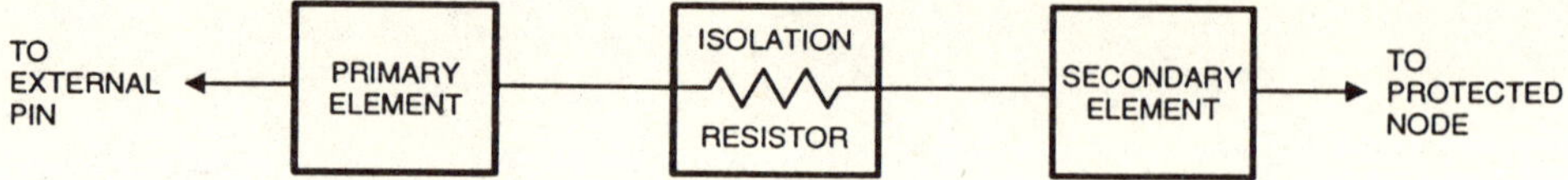

Figure 4.1 Conventional Input Protection Scheme. The primary and the secondary protection devices are separated by an isolation resistor.

technology. Next, the chip function must be considered and the operating requirements for the circuit being protected must be compatible with the choice of protection circuit. This includes the operating voltage conditions, capacitance and resistance loading, and area requirements. Finally, the customer and product engineering requirements for ESD and handling capability must be weighed against the possible impact of the protection circuit on the performance of the IC. For the protection circuit designs described here, each of these aspects is considered in detail.

A common protection circuit schematic is composed of a primary element and a secondary element as shown in Figure 4.1. The primary element will shunt most or all of the current during an ESD event, while the secondary element serves to limit the voltage or current at the circuit being protected until the primary device is fully operational. The two elements are isolated by a resistive element. There are several candidates for a primary protection device. It can be a thick field transistor, a Silicon Controlled Rectifier (SCR), an nMOS transistor, or a simple *pn* diode. In this chapter we will discuss some of the commonly used protection devices. In each case, the main features, the advantages, proper methods for layout, and practical limitations are discussed.

The effectiveness of the primary device in most cases is determined by the secondary protection stage. The secondary device can be a small grounded gate MOS transistor or a diode between the pads and the power/ground supplies. The overall design critically depends on the choice of these devices along with proper selection of the resistor element. The resistor element can be polysilicon, n^+ diffusion, p^+ diffusion, or n-well. If a zener diode is to be used as a secondary device in a standard CMOS process, then special process steps are needed to build the diode.

The protection devices mentioned above are described in the following sections. The effectiveness of the total protection concept is realized only when all of the components are harmonized to work together during an ESD event. To achieve this, a design synthesis approach is needed as illustrated in Section 4.4.

4.2.1 Thick field device

The thick oxide or field oxide device (FOD) has been commonly used as a protection circuit element for technologies with feature sizes (defined by the nominal polysilicon gate length) ranging from 3 μm to 1 μm. The standard cross-section of this device is shown in Figure 4.2. In effect, the FOD operates as a lateral bipolar transistor as described in Chapter 3. The spacing from the drain contact to the

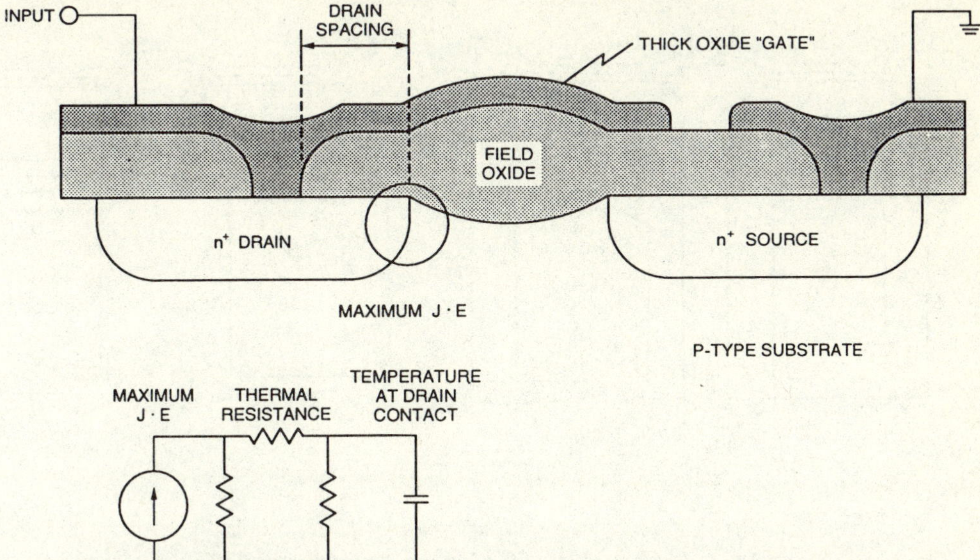

Figure 4.2   Cross-section of a thick field oxide device. The maximum heat occurs due to $J \cdot E$ at the cylindrical junction. The electrical analog is shown in the bottom of the figure.

diffusion edge or *drain spacing* is a critical parameter for improved ESD. As indicated in the figure, the maximum heating during ESD occurs at the cylindrical junction where $J \cdot E$ is maximum. The FOD can be laid out in different ways for use as an ESD protection circuit. Three different cross-sections of this device are shown in Figure 4.3. In Figure 4.3(A), the device is formed with no gate. Since the onset of bipolar action is determined by n^+ drain diffusion breakdown, a gate is not necessary for *npn* snapback to take place. In Figure 4.3(B), the same device is shown with a metal gate connected to the drain pad. In this case, as the voltage at the drain pad increases, the metal gate is intended to reduce the source (emitter) barrier and turn on the *npn* device at a lower level [Duvvury83]. The lowest trigger voltage can be obtained using a polysilicon gate as shown in Figure 4.3(C). Extensive studies have shown that the metal gate connection has no obvious effect on the ESD failure threshold [Wilson87][DeChiaro86]. There are no known data reported on a polysilicon gate structure. In general, the bipolar trigger voltage of the FOD is determined by the n^+ to n^+ spacing as shown by the experimental results in Figure 4.4 [Weston92].

The main design parameters of the FOD as far as ESD is concerned are the channel length (L), the drain contact-to-diffusion spacing towards the channel (DS), and the device width (W). These three parameters are identified in Figure 4.5. First consider the channel length dependence. It has been shown [Palella85] that the HBM failure thresholds increase as the channel length is decreased in the 7 μm to 2 μm range. In contrast, DeChiaro [DeChiaro86] reported that the thresholds increase with increasing channel length in the range of 1 μm to 3 μm. Work by Rountree and Hutchins [Rountree85] and Duvvury, Rowntree and White [Duvvury83] showed that

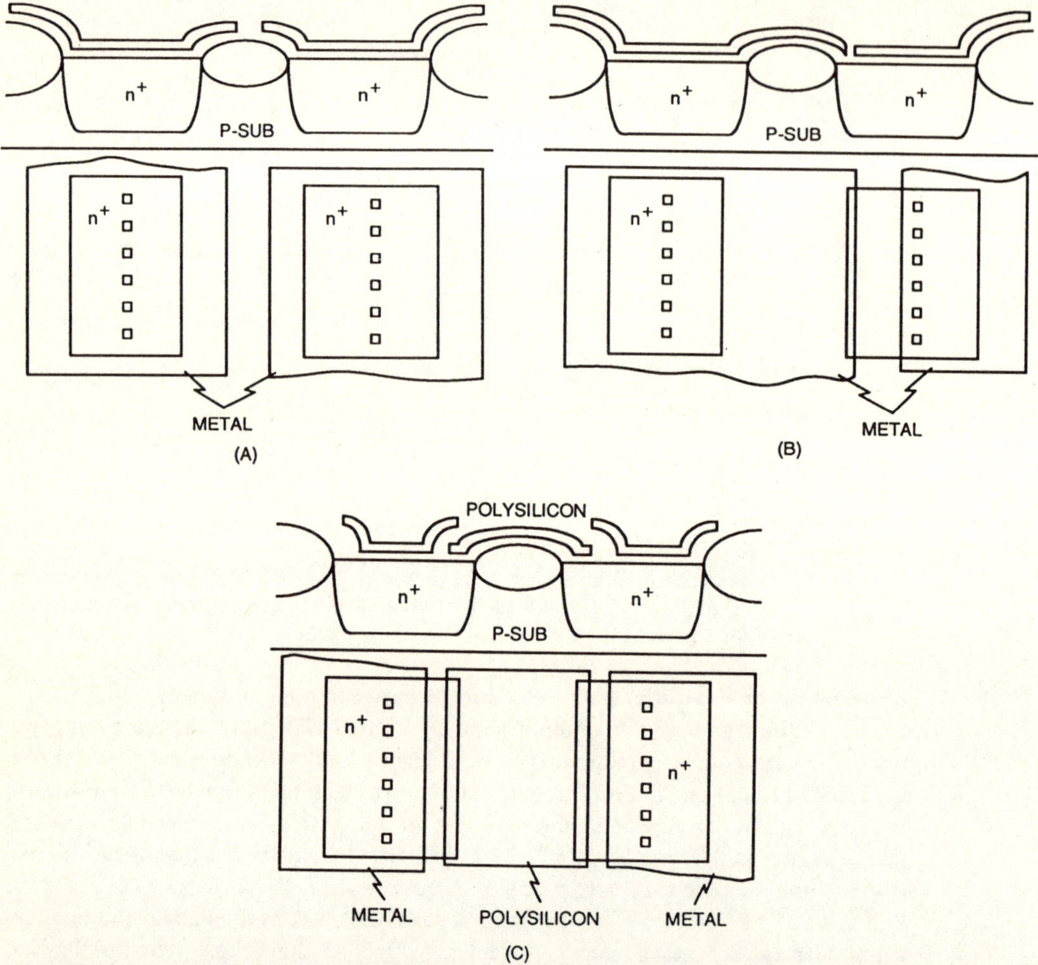

Figure 4.3 Various layout styles for the thick oxide protection device. Structure (A) has no gate, structure (B) has a metal gate, and structure (C) has a polysilicon gate.

there was little dependence on channel length between 2 μm to 8 μm. These different conclusions are probably due to the different technologies and structures used in these studies. A detailed investigation of this phenomenon [Wilson87] concluded that all three different channel length dependencies might appear if the data is carefully collected over a wide range of dimensions. Hence, for a given technology there might very well exist an optimum point close to, but not at, the minimum allowed channel length. As a general rule, for a 2 μm technology the optimum point could be as low as 1 μm or as high as 3 μm. It could be argued that at the minimum point punchthrough might dominate, leading to some reduction in the failure threshold, and at longer lengths the bipolar efficiency goes down, also reducing the failure thresholds.

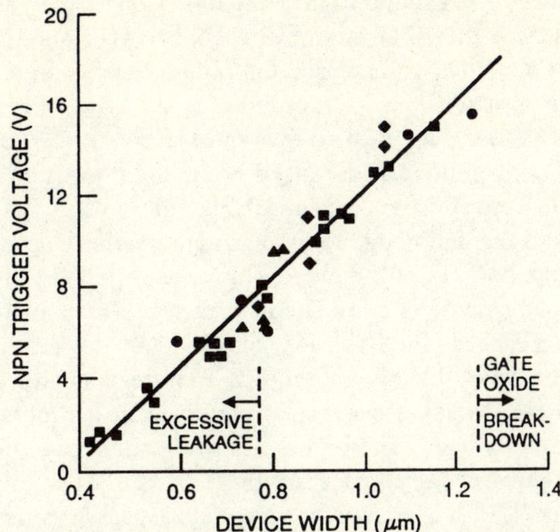

Figure 4.4 The trigger voltage of the thick oxide device as a function of field oxide width after Weston *et al* [Weston92]. Note that the 'width' in this figure actually represents the channel length of the device, parameter L in Figure 4.5.

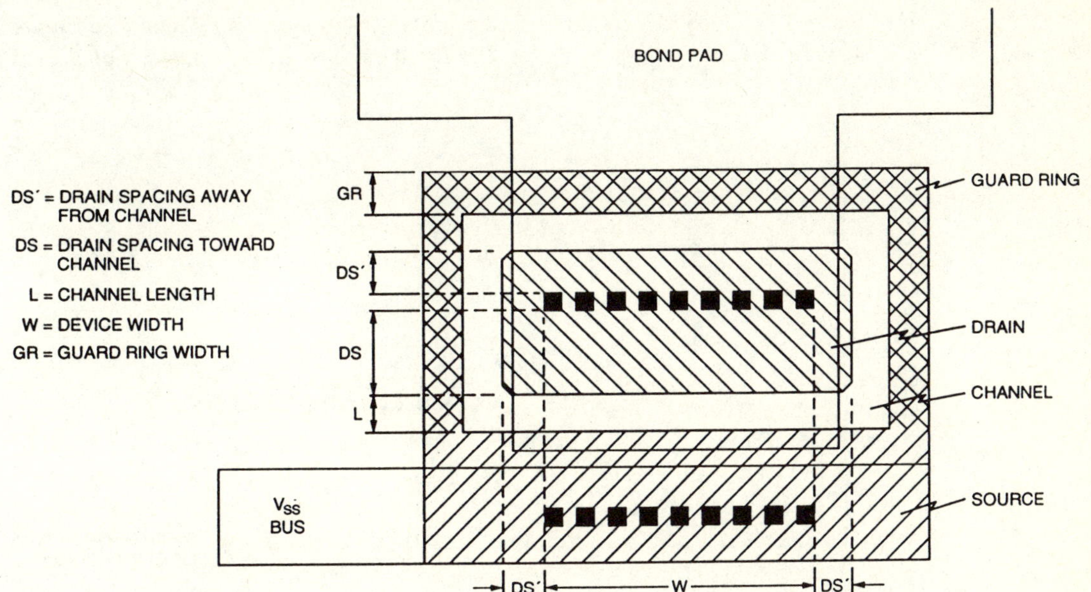

Figure 4.5 Layout for a thick field device with all the critical ESD design parameters.

The second parameter to be considered is the drain contact to gate spacing. The original work by Rountree and Hutchins [Rountree85] showed that the impact of this parameter on the ESD failure threshold is very significant. This data is shown in Figure 4.6. As *DS* is increased the failure level is seen to increase to a certain level and then saturate. It was proposed that the reason for the influence of *DS* was the distance between the heat source and the contact. When the contact is close to the diffusion edge, the heat produced at the drain junction isotropically spreads to heat the contact metalization and result in a lower failure voltage. Moving this contact away from the drain edge to an optimum amount improves the failure level. Further increasing *DS* will only have an incremental effect on ESD performance once the weak failure mode is eliminated. In fact, as *DS* is increased even further the extra series resistance can degrade the bipolar device performance and lower the protection level. In this extreme case, instead of the avalanche breakdown taking place at the diffusion edge or at the sidewall it occurs at the bottom wall directly beneath the contact. The bipolar does not trigger and damage occurs as a result of spiking between the contact and the substrate. This failure mechanism is illustrated in Figure 4.7. A technique where large spacing could be used for its advantage without contact spiking is to place *n*-well directly under the drain contacts whereby the bottom wall junction avalanche voltage is made substantially higher than the avalanche voltage at the cylindrical sidewall.

The contact spacing phenomenon has been investigated by several other workers (see e.g. [McPhee86][Wilson87][Palella85]) and all reached similar conclusions for a non-silicided process. The contact spacing away from the channel (indicated as *DS'* in Figure 4.5) was found to have no impact on the failure level.

Although the impact of the contact spacing had been very important for the abrupt junction processes, its effect was found to be weak for LDD junctions [McPhee86][Duvvury85]. As a result, the failure voltage versus *DS* curve in Figure 4.8 shows a limiting effect at larger *DS* for the 1 μm LDD process. For a clad

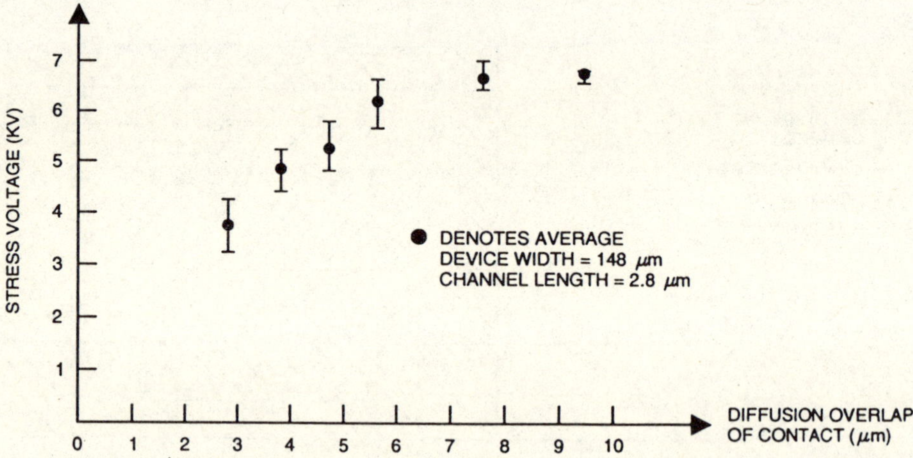

Figure 4.6   Failure voltage for the HBM ESD stress versus the drain contact to diffusion edge spacing, after Rountree and Hutchins [Rountree85].

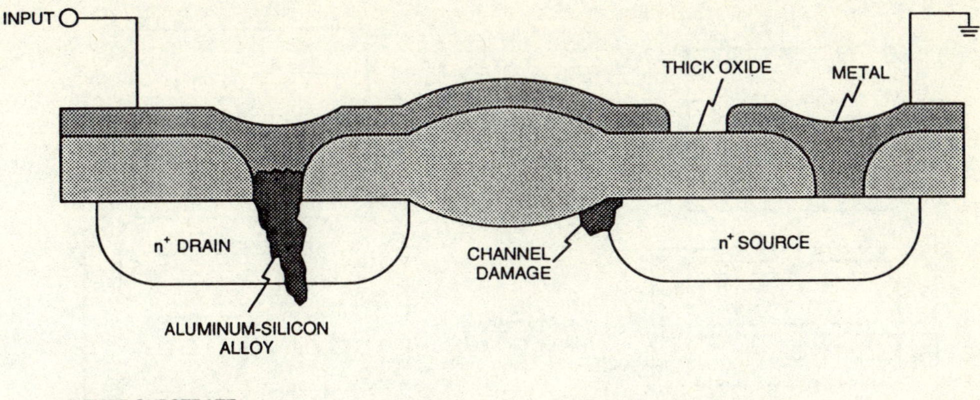

Figure 4.7 Cross-section of a thick field oxide device with two different failure modes.

silicide process, however, the effect of the drain contact spacing virtually vanishes [McPhee86][Duvvury86][Wilson87]. In all of these works it was clearly shown that the clad silicide drain diffusion reduces the resistance of the drain contact to gate region and eliminates the ballasting that is critical for good ESD levels. This typical result is illustrated in Figure 4.8. Process changes to the silicide can be used to improve ESD levels. For example, a thinner silicided layer with its relatively increased resistance can result in some improvement in ESD performance [Chen88][Duvvury89]. However, the most effective solution is to block the silicide formation in the region between the drain contact and the gate (or diffusion) edge although this increases process complexity and is an expensive option.

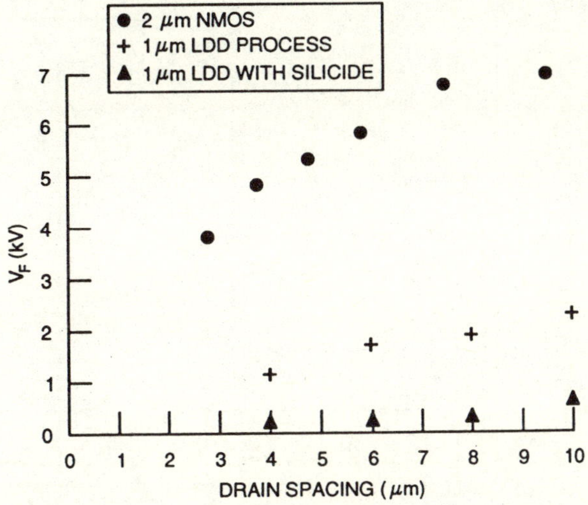

Figure 4.8 Failure voltage for the HBM ESD stress versus the drain contact to diffusion edge spacing for three different process options.

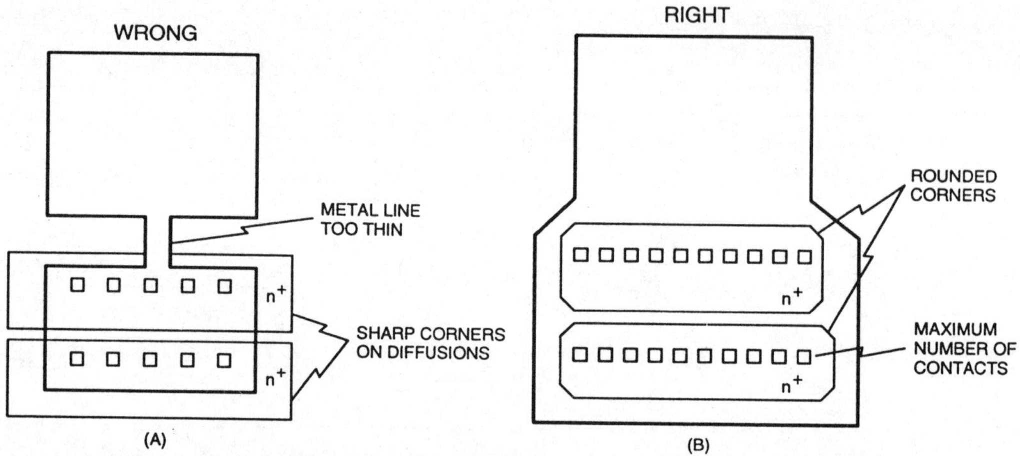

Figure 4.9   Layout styles for a thick field device showing (A) the 'wrong' and (B) the recommended 'right' approach.

Besides the drain contact spacing, the overall layout of an ESD protection device is critical. To illustrate this for the thick field oxide device, the *wrong* and *right* layout approaches are shown in Figure 4.9(A) and Figure 4.9(B), respectively. In this layout the spacing between contacts is uniform and the contacts are equidistant from the diffusion edge which is essential for ensuring uniform current flow across the device. Current uniformity reduces thermal effects by spreading out the heat dissipation. The diffusion edges are shown to be rounded which reduces the electric field effects at the corners and prevents the localization of current flow in these areas. The overall protection scheme with a thick field oxide device is shown in Figure

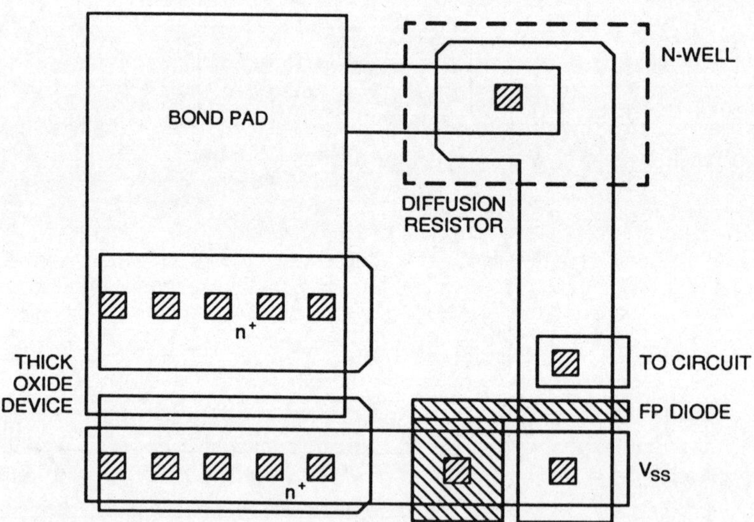

Figure 4.10   Layout for a total input protection with a thick oxide device, a diffusion isolation resistor, and a field plated diode.

4.10. Note that this layout corresponds to the circuit schematic in Figure 4.1. In this figure FP DIODE is the field plated diode which is a small grounded gate nMOS that will clamp the voltage during the ESD pulse below the breakdown voltage level of the input gate oxide. After the lateral *npn* transistor associated with the grounded-gate nMOS turns on, the current through the diffusion resistor increases the pad voltage and triggers the lateral *npn* transistor associated with the FOD. Embedding the diffusion resistor head in an *n*-well helps to suppress avalanche breakdown between the n^+ diffusion and the *p*-substrate and improves the ESD robustness.

4.2.2 nMOS transistors (FPDs)

nMOS transistors are essentially thin oxide devices as opposed to the FOD discussed in the previous section. They are also called field-plated diodes (FPD) or gated-diodes in the literature because of the effect of the gate on the diode breakdown voltage. The FPD has not been commonly used as a primary protection device, in comparison to the FOD in technologies with feature sizes >1 μm. The reason for this is that the FPD requires a relatively larger device size to achieve the equivalent level of protection as an FOD. However, in advanced processes with LDD junctions the FOD performance is limited and the onset of damage has been observed at between 1 kV and 2 kV. Hence, nMOS devices are gaining more usage as primary ESD protection devices in advanced CMOS processes.

For larger feature size technologies, the FOD is better since the bipolar action takes place deeper in the silicon and the peak heating is located further away from the silicon surface. In nMOS devices the peak heating occurs close to the surface which has a poor thermal conductivity. The result is damage to the silicon surface and the formation of melt filaments between the polysilicon gate and the silicon surface. A typical failure site is shown in Figure 4.11.

While the FOD can give as much as 40 V of ESD performance per μm of device width for the abrupt junction processes, the performance of the FPD per unit width is relatively lower. A comparison between the two types of devices is shown in Figure 4.12, for non-silicided technologies. It is seen that with the introduction of LDD junctions at 1.5 μm, intended to improve the hot carrier reliability, a significant drop in the performance occurred for both the FOD and the FPD. But in the submicron range the thin oxide device actually gives a slightly better performance. Analysis has indicated that as technologies approach the 0.10 μm regime the ESD capability of the nMOS device will be better than in the 1 μm technologies [Lin93]. This is attributed to the decreased power dissipation in these devices as the avalanche breakdown voltage and snapback holding voltages are reduced in the scaled technologies. It has been shown [Amerasekera94] that this turnaround in the ESD performance of the nMOS device occurs at the 0.8 μm technology node for non-silicided devices and at the 1 μm technology node for silicided devices as shown in Figure 4.13.

The main design parameters of the nMOS transistor as shown in Figure 4.14 are the transistor channel length (L), the drain contact-to-gate spacing (DCG), and the device width (not shown). The source contact-to-gate spacing (SCG) does not play

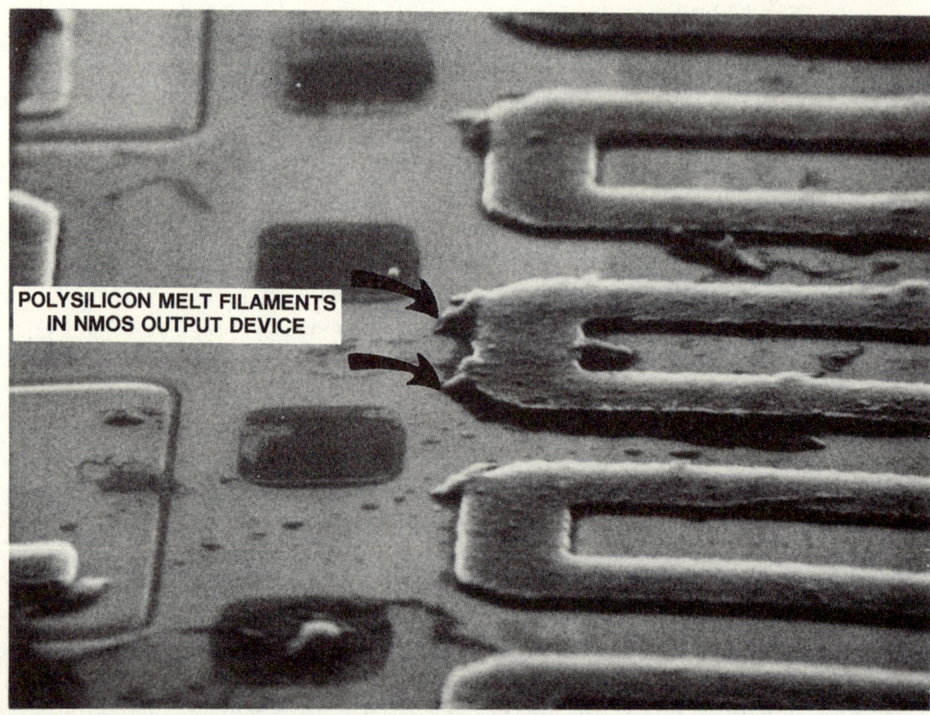

Figure 4.11 Typical failure site in nMOS output device showing melt filaments between polysilicon gate and silicon surface.

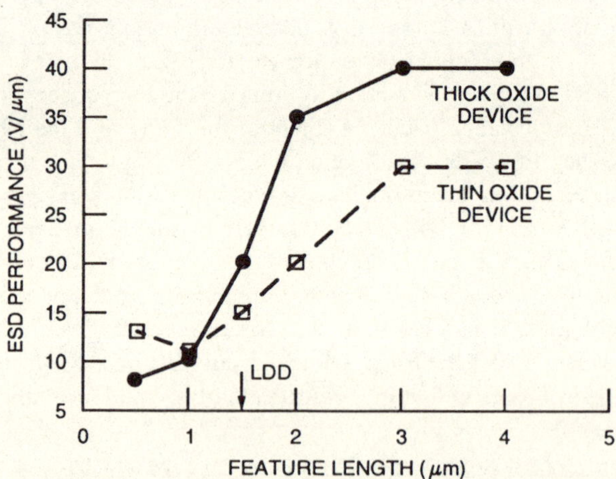

Figure 4.12 The ESD performance in V/μm as a function of feature size for thick oxide and thin oxide non-silicided devices.

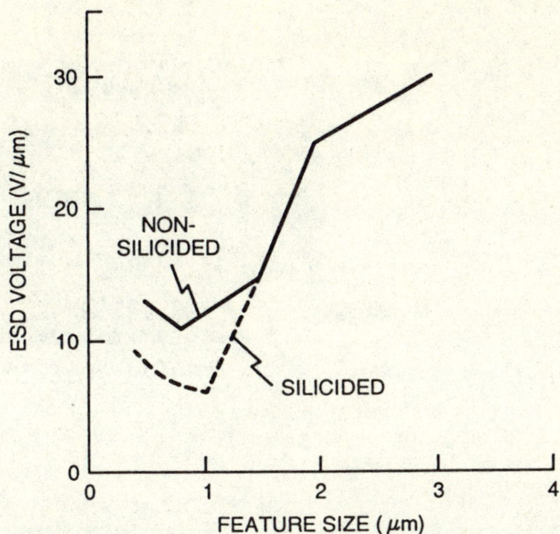

Figure 4.13 The ESD performance in V/μm as a function of feature size for a thin oxide devices, compared for both silicided and non-silicided processes.

much of a role and is often kept at its minimum design value. Since a large device width is required to obtain a given ESD level, the device is typically laid out as a finger or ladder structure as shown in Figure 4.15. This type of layout is often used for output transistors and provides current uniformity between fingers and maximizes the ESD protection level. The parasitic resistance R_F is composed of the metal finger resistance, contact resistance, and diffusion sheet resistance. The parasitic resistance R_S is that associated with the ground bus that connects the fingers of the device. Maintaining a minimum ratio for R_S/R_F is important for obtaining the best possible ESD performance. Minimizing R_S will give more uniform current distribution as shown in Figure 4.16 [Duvvury88]. Note that while maximum R_F is desirable, the contact resistance itself should not be increased since damage could then result from excessive contact heating. For the protection device layout shown in Figure 4.15, the nMOS transistor has its gate tied to ground. However, multifinger structures have been found to have inconsistent performance as ESD protection devices because of non-uniformity of the finger turn-on during an ESD event [Chen88][Polgreen89]. In this respect the substrate connection is important. For

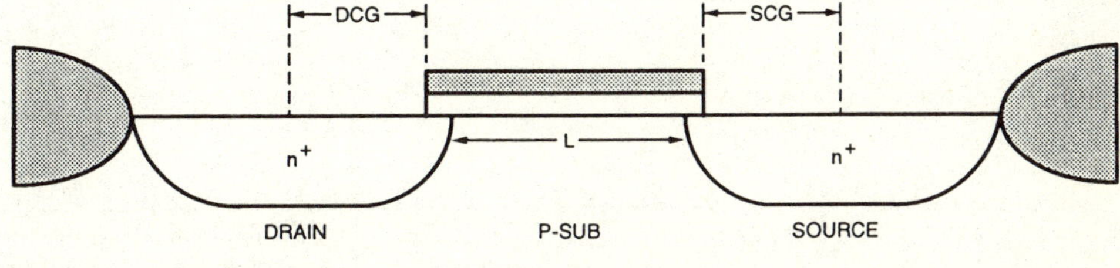

Figure 4.14 Typical ESD failure mode in an nMOS transistor showing gate to drain melt filaments, showing critical ESD design parameters.

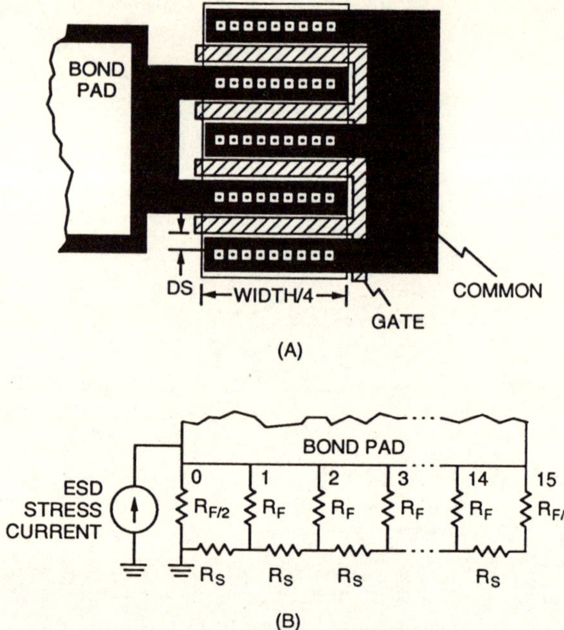

Figure 4.15 The ladder type layout for an nMOS transistor and the equivalent electrical representation of the parasitics.

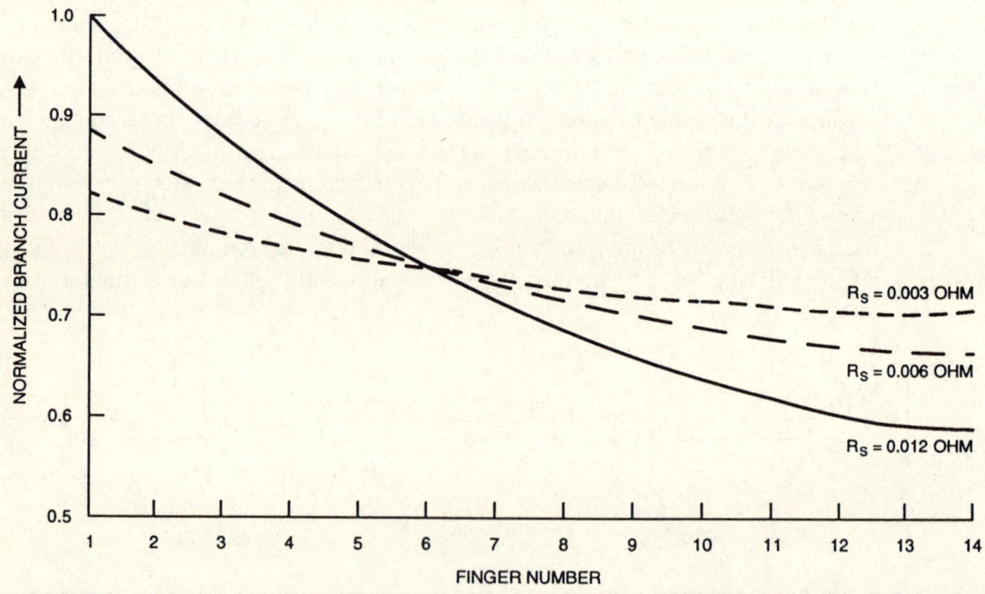

Figure 4.16 Simulated branch currents in the fingers of an nMOS transistor for different resistance values for the source bus.

floating substrate devices such as DRAMs, all the fingers turn on effectively allowing maximum performance from the device.

The floating substrate nMOS device is essentially a floating base *npn* device. In Chapter 3 we explained how the base resistance influences the *npn* turn-on voltage and the floating-base *npn* has a much lower trigger voltage than a grounded-base device. Thus the difference between the trigger voltage and the on-voltage of the *npn* is small and it is possible to turn on more fingers before the damage threshold of the device is reached.

For logic applications the substrate is tied to ground along with the source and so the uniform turn-on for the device cannot be obtained through modulation of the base resistance. However, as described in recent work [Duvvury92], the gate can be coupled high during the ESD event allowing MOS current conduction. As described in Chapter 3 this has the effect of lowering the *npn* turn-on voltage which again increases the number of fingers that are effective during an ESD event. Therefore, the device gate modulation is another important design parameter for the nMOS protection device. This device is described in more detail in Section 4.2.3.

The channel length of the thin oxide device does play a role in its ESD protection capability. Similar to the thick oxide device, a minimum channel length is desired for efficient turn-on but the punch through limit and the associated leakage should be avoided. This is especially important when using the output buffer device as the protection device itself. The minimum channel length device is also known to have higher susceptibility to hot carrier stress which means that the channel length needs to be optimized between the ESD performance requirement and hot carrier requirement.

The ESD failure threshold of a 200 μm thin oxide device as a function of the drain contact to gate spacing for different technologies is shown in Figure 4.17. As expected, the drain contact spacing has a large effect for non-silicided processes. It is interesting to note that for the three non-silicided technologies evaluated in this work, the optimum spacing is ≈ 6 μm. For the silicided cases there is no obvious dependence on this parameter. However, one study [Duvvury90] did find that a minimum spacing gives a better distribution. This result is shown in Figure 4.18 where the distributions are shown for silicided and non-silicided devices. Even then, the nMOS protection device is still very inefficient in a silicided process.

In contrast to the drain side contact spacing, the source side spacing is not important for grounded substrate technologies. This is because, for a negative voltage applied to the drain, the n^+ to p-substrate diode operates in the forward-biased condition and there is no current flow or heat dissipation in the source diffusion. Keeping the source contact spacing at a minimum is the best approach for minimizing the power dissipation when in the *npn* mode, as well as for reducing the area of the protection device. For floating substrate technologies there is no forward biased diode for negative applied voltage stress. Under these conditions the *npn* is triggered for both positive and negative stress polarities. Since the stress condition is symmetrical, the same 6 μm spacing should be maintained at both source and drain diffusions.

In conclusion, the thin oxide device can provide good ESD protection levels. A typical design would have a transistor width of ≈ 200 μm with close to minimum

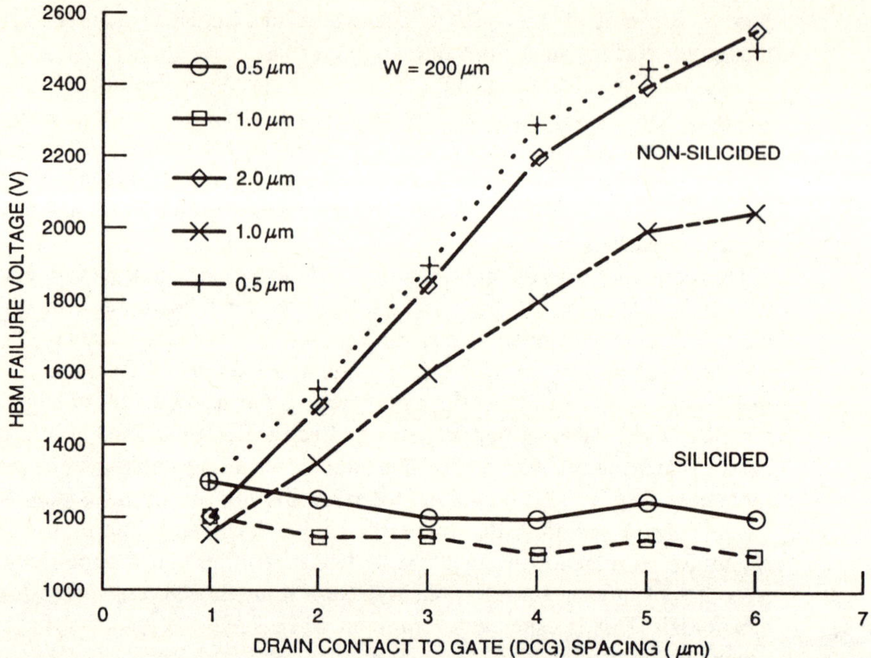

Figure 4.17 The HBM ESD performance in thin oxide transistors as a function of drain contact to gate spacing. The different silicided and non-silicided processes are as indicated.

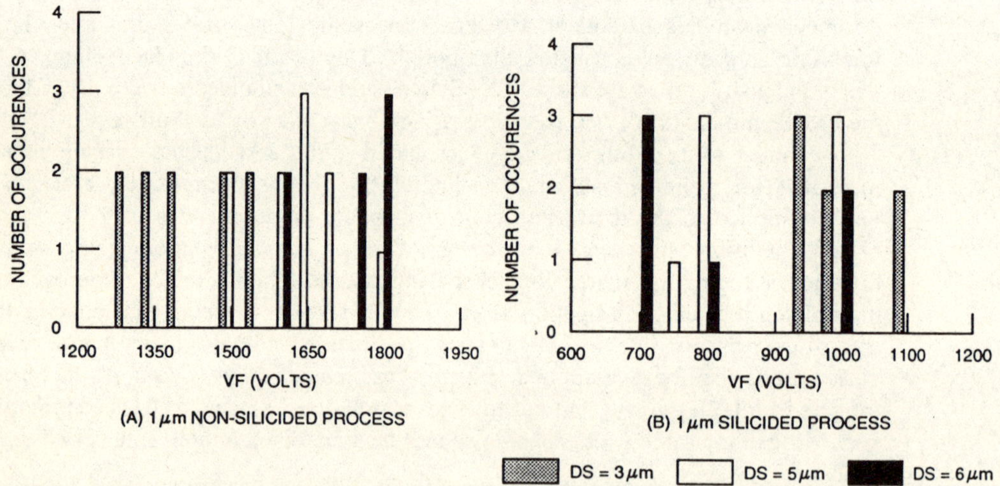

Figure 4.18 HBM ESD failure distributions as a function of drain contact spacings for non-silicided processes (A) and for silicided processes (B).

channel length. The drain contact to gate spacing for a non-silicided process should be about 6 μm while the minimum allowed spacing should be used for a silicided process. The spacing on the source side can be kept at a minimum. For a multifinger structure in a non-silicided process, the use of more than two or three fingers will not really serve to increase the protection level unless design techniques such as the gate coupling method are used to improve this efficiency. Note also that the finger length will play some role. For excessively long fingers the voltage dropped within the finger will not make it effective and the heating tends to localize in the middle of the finger [Scott86]. This is especially true for silicided processes. As a good practice, 40–80 μm finger lengths are recommended.

4.2.3 Gate-coupled nMOS (GCNMOS)

In most applications, the thin oxide device is used as a protection device with its gate grounded. This will always ensure that the protection device, while being effective for ESD protection, will not cause any extra leakage at the pin. However, the thin oxide device can be a more effective protection element if its gate is biased high during an ESD event. The effect of gate bias on the ESD performance of nMOS devices has been reported [Chen88][Polgreen89][Abderhalden91]. As illustrated in Figure 4.19, if the gate voltage is about 1 V the nMOS trigger V'_{t1} is lowered to less than the voltage required for the onset of second breakdown V_{t2} and is therefore ideal for improved ESD protection. It was also noted that if the gate voltage goes above 5 V, the second breakdown trigger current I_{t2} (defined in Figure

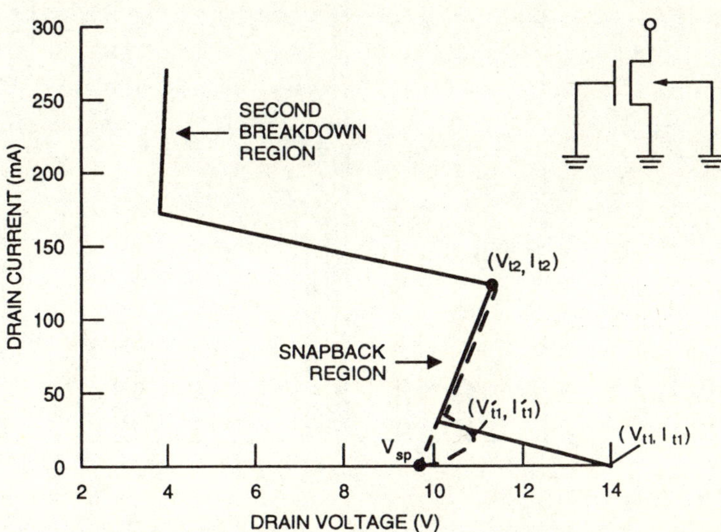

Figure 4.19   The high current *I–V* curve for an nMOS transistor. The solid line is for the case with the gate grounded and the dashed line is for the case with the gate coupled above V_t.

4.19) decreases to give reduced failure threshold voltage ($I_{t2'} < I_{t2}$). Hence, a gate voltage of between 1 V and 2 V is typically needed for best ESD performance.

In a multifinger structure, the gate coupling improves the uniform turn-on of all the fingers. This is not always possible in a grounded gate device since the first *npn* device to turn on discharges all the ESD current, potentially preventing the other fingers from turning on. In such a device, each of the other fingers has a chance to turn on as the voltage increases again towards V_{t2}. That is, after one finger begins *npn* conduction and clamps at the snapback voltage, the pad voltage builds up again due to the snapback resistance. When the pad voltage again reaches V_{t1} the next finger turns on, and so on until all the fingers are turned on, or the failure current I_{t2} is reached, whichever comes first. Usually I_{t2} (or more accurately V_{t2}) is reached first and the total number of fingers that actually turn on varies substantially as shown by the failure distribution in Figure 4.20. Now the effect of gate coupling on the transistor threshold voltage is to lower the avalanche breakdown voltage, V_{br}, as shown in Figure 4.21. This is shown here for four different technologies. Note that in all cases the minimum V_{br} (same as V_{t1} in Figure 4.19) is reached when the gate is between 1 V and 2 V.

There are different techniques to achieve gate coupling during an ESD event. In the structure shown in Figure 4.22, the coupling on the gate is determined by the ratio of the gate-drain overlap capacitance to the thin oxide capacitance [Duvvury92]. Typically, after the *npn* turns on and clamps the voltage at ~ 8 V,

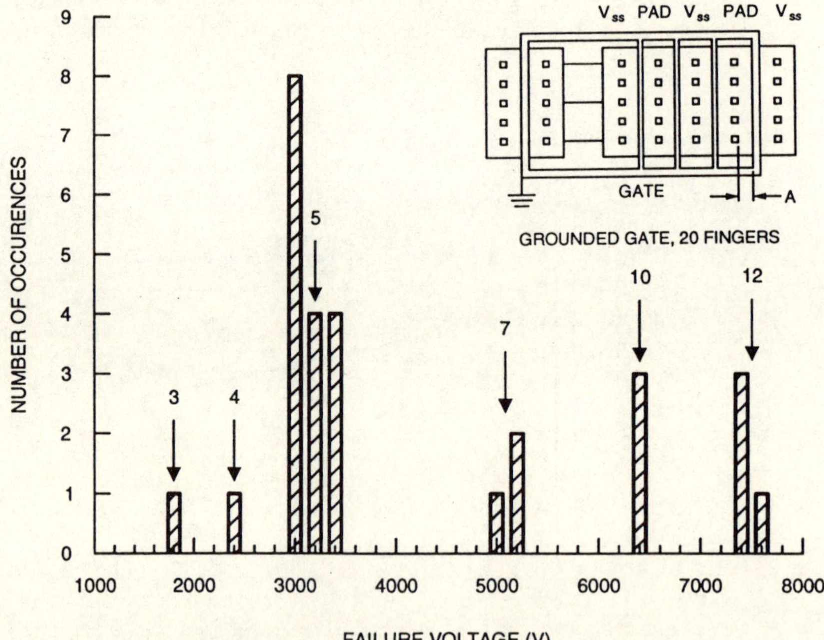

Figure 4.20 The ESD failure distribution for a grounded gate nMOS transistor. The numbers indicated are the number of fingers assumed to be turned on during ESD.

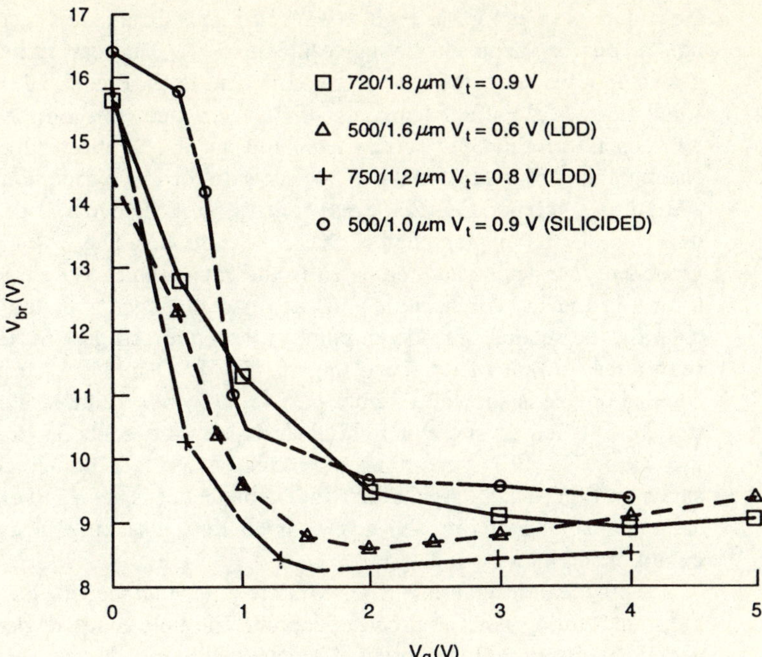

Figure 4.21 The avalanche breakdown voltage as a function of gate bias in an nMOS transistor for four different technologies.

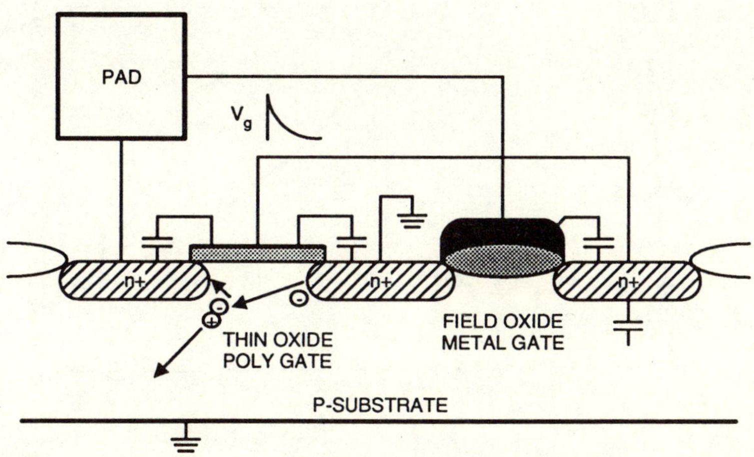

Figure 4.22 Cross-section of a gate-coupled nMOS (GCNMOS) where its gate is connected to ground through a thick oxide device.

the pad voltage can go as high as 15 V for ESD currents of around 2 A due to the bipolar device snapback resistance (about 5 Ω). This can turn on the field oxide device and discharge the gate potential to zero level. Figure 4.23 shows circuit level simulations of these effects using SPICE. The time constant for the gate discharge will depend on the ESD current level and the size of the field oxide device width. Usually it is designed so that the gate stays on for a minimum duration of between 5 and 10 ns corresponding to the rise time of the ESD event. This will allow enough time for all the fingers in the nMOS to turn on. The $I–V$ characteristics of a grounded gate device are compared to the gate-coupled device in Figure 4.24. The arrows in the inset indicate where each nMOS finger goes into *npn* snapback. In contrast, the second breakdown point in the grounded gate devices is reached very soon after turn-on of only two fingers. The ESD failure distribution of the gate-coupled device is shown in Figure 4.25 for a 1 μm silicided technology. The failure threshold voltages show a tight distribution and scale as the device width is increased. The ESD performance for this device in a non-silicided technology is shown in Figure 4.26. Again the width dependence is seen and excellent ESD levels are obtained. Therefore, the gate-coupled device is an effective ESD protection circuit element.

The practical design of the GCNMOS requires that the gate has a good connection to ground during normal circuit operation. The field oxide device at the gate of the nMOS by itself will not bring the gate to ground during normal operation and

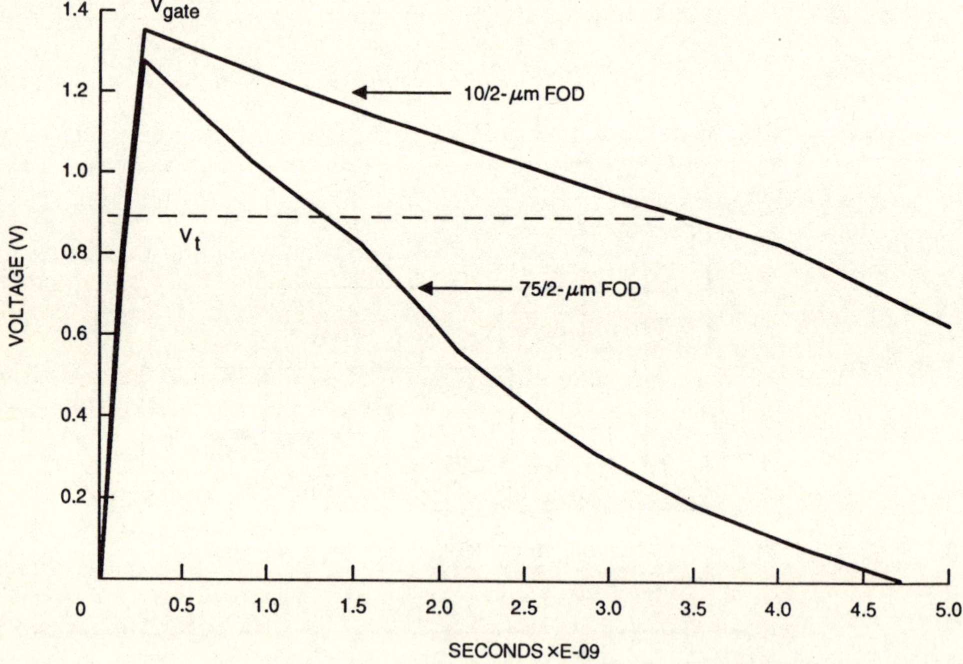

Figure 4.23 Simulated gate transient waveforms for a GCNMOS. The size of the thick field oxide (FOD) has an influence on the gate decay as shown here.

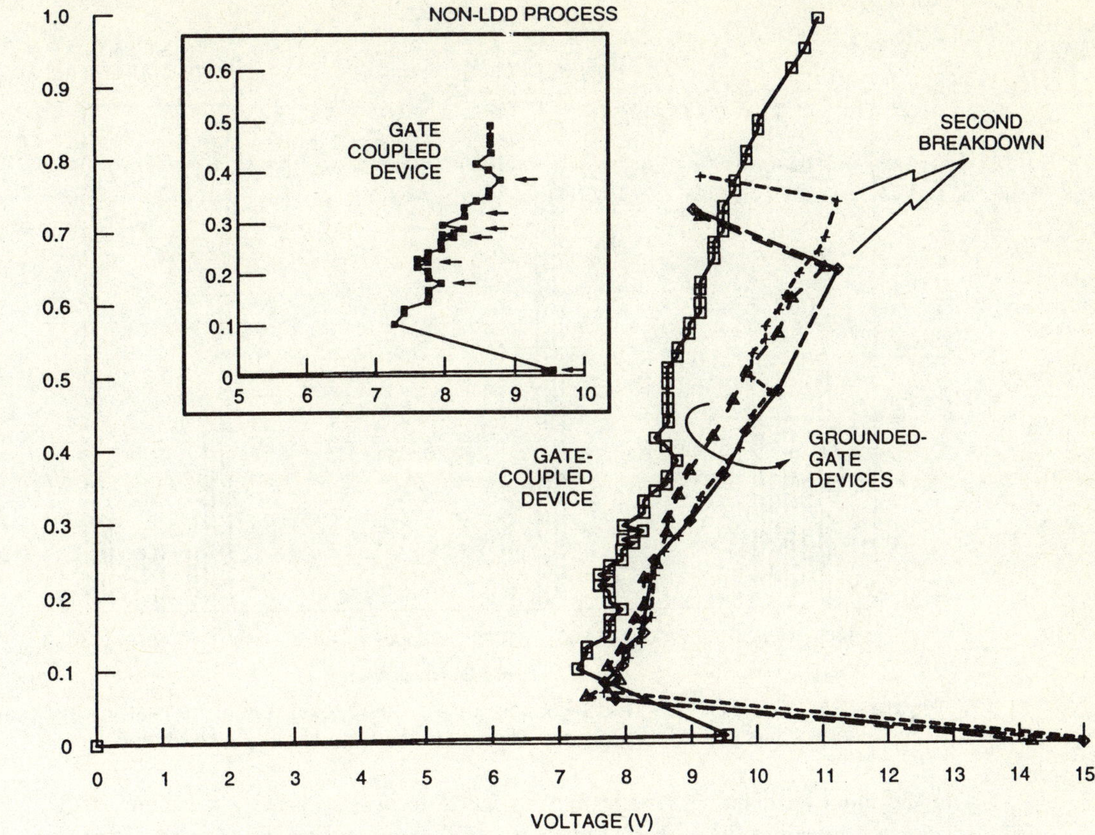

Figure 4.24 The measured $I-V$ curves for a gate-coupled nMOS compared to a grounded gate nMOS. The inset shows the details of individual finger turn-on for the gate-coupled device.

unacceptable leakage will result if the gate remains floating. A more effective approach is to connect the gate to ground through a large resistor (≈10–15 kΩ) so that the gate voltage is discharged by the time constant determined by this resistance and the total capacitance at the gate. Other techniques using external circuitry to raise the gate voltage during an ESD event have also been reported [Krakauer94].

4.2.4 SCR protection devices

The Silicon Controlled Rectifier (or SCR) is the most efficient of all protection devices in terms of ESD performance per unit area. The basic SCR is a *pnpn* device as shown in Figure 4.27. The device shown in Figure 4.27 is also referred to as a lateral SCR or LSCR.

The operation of the SCR has been described in Chapter 3. We will briefly repeat the main points here. The adjacent n^+ and p^+ diffusions in the n-well are connected

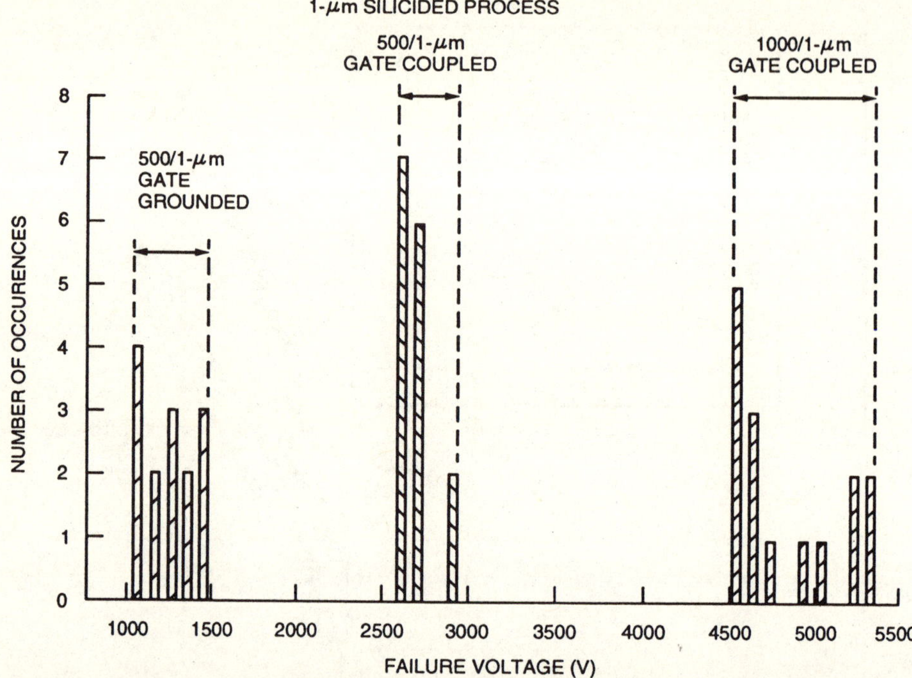

Figure 4.25   The ESD failure distribution for a gate-coupled nMOS transistor for a 1 μm silicided process. The grounded gate case is also shown.

to the input terminal. A vertical *pnp* device is formed with the *p*-substrate as the collector, *n*-well as the base, and input p^+ diffusion as the emitter. The n^+ diffusion in the *p*-well is connected to the ground or substrate bus and forms the emitter of the *npn* transistor. The base of the *npn* is formed by the *p*-substrate and the collector is the *n*-well and the *n*-well contact. During normal circuit operation, CMOS latchup should not be a problem since the emitter and base of the *pnp* are at the same potential. During an ESD stress pulse the collector-base junction of the *npn* goes into avalanche breakdown generating the electron current in the *n*-well which forward-biases the emitter-base junction of the *pnp*. The turn-on of the *pnp* occurs in less than 1 ns and this leads to the regenerative *pnpn* action [Rountree88]. Once the SCR is turned on the device is in a low impedance state and the anode to cathode clamping voltage is of the order of 1 V to 2 V in a submicron process. This dramatically reduces the power dissipation and results in an improved ESD performance. The nature of the device operation means that it is not strongly influenced by salicidation which is a big advantage in advanced CMOS processes. The performance of the device is higher (60–70 V/μm) in non-salicided processes than in salicided processes (40–50 V/μm) but the performance is so high that the difference is not that important. Other key process-controlled parameters for successful operation of this device are the holding voltage controlled by the *n*-well overlap of the p^+ anode (X), and the trigger voltage determined by the *p*-substrate resistance, R_p in Figure 4.27. Therefore, a thicker epitaxial layer is desirable for

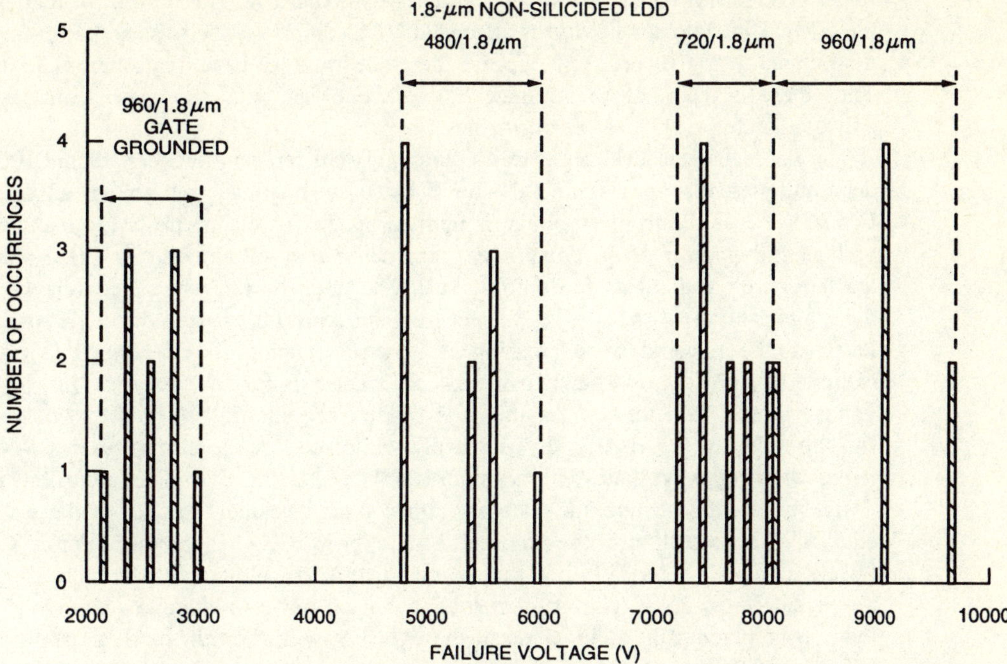

Figure 4.26   The ESD failure distribution for a gate-coupled nMOS transistor for a 1.8 μm non-silicided process. The grounded gate case is also shown.

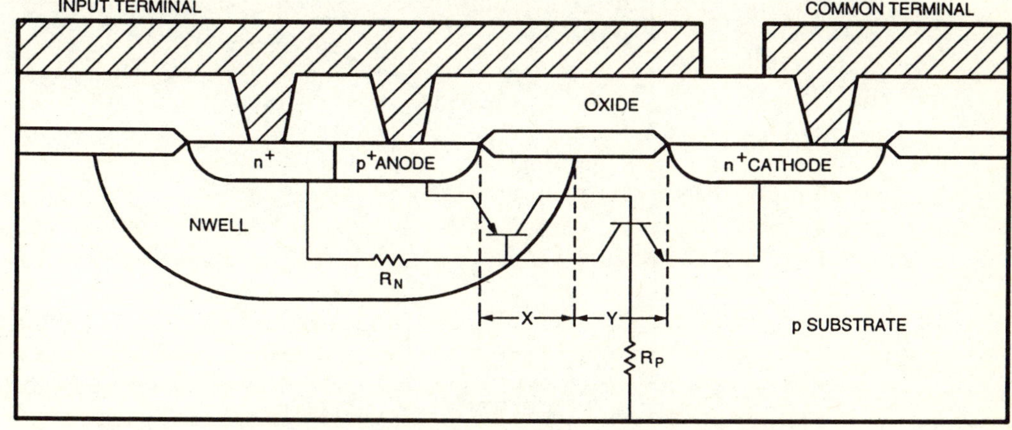

Figure 4.27   Cross-section of a lateral SCR device.

better ESD performance. However, a thinner epitaxial layer is required for reducing the CMOS latchup sensitivity in advanced VLSI chips. In such cases the *npn* may not be able to trigger properly because the *p*-substrate (or base) resistance is so low. Thus there is the need to optimize the choice of epitaxial thickness to trade-off between latchup and ESD performance.

The LSCR trigger level is generally quite high and can vary between 40 and 100 V depending on the process and design [Chatterjee91A][Rieck89]. In an advanced CMOS process the trigger voltage is defined by the *n*-well to substrate breakdown voltage and is about 50 V. The process parameters that influence the trigger voltage are the *n*-well and substrate doping levels. The main design parameter influencing the trigger voltage is the spacing between the anode and the *n*-well edge. The trigger level can be lowered by decreasing the critical spacing for the anode (*X*) and cathode (*Y*) but this can lead to increased leakage. Figure 4.28 shows the trigger voltage plotted as a function of the SCR spacings. It should be noted that below 3 μm the SCR trigger voltage drops sharply while the leakage current increases. To reduce the trigger voltage without significantly impacting the leakage current, the design is modified to include a highly doped region near the surface at the *n*-well edge. The cross section of the modified LSCR (or MLSCR) is shown in Figure 4.29 (inset) along with its *I–V* breakdown characteristics. Note that the SCR trigger now occurs at ≈ 25 V. This trigger voltage can be further reduced to the 12–15 V range by replacing the field oxide of the MLSCR with the thin oxide as shown in Figure 4.30 [Chatterjee91A]. Because of its low trigger level, this last device is called a Low Voltage Trigger SCR or LVTSCR. The LVTSCR essentially uses a MOS device in parallel with the SCR. Triggering occurs after the drain junction of the MOS transistor begins avalanching. The avalanche-generated hole current in the *p*-substrate turns on the lateral *npn* and then the vertical *pnp* followed by eventual regenerative SCR action. The low trigger voltage of the LVTSCR means that it can be used as an ESD protection device for CMOS output buffers. However, one still needs to ensure that the output device does not trigger before the SCR. This can be

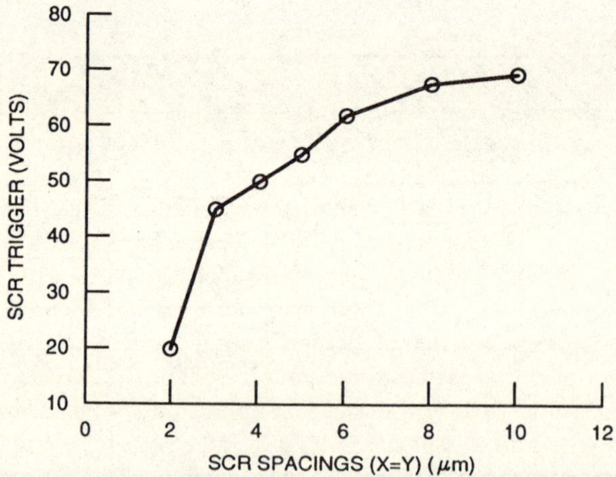

Figure 4.28   The SCR trigger voltage as a function of SCR spacing (*X* = *Y*).

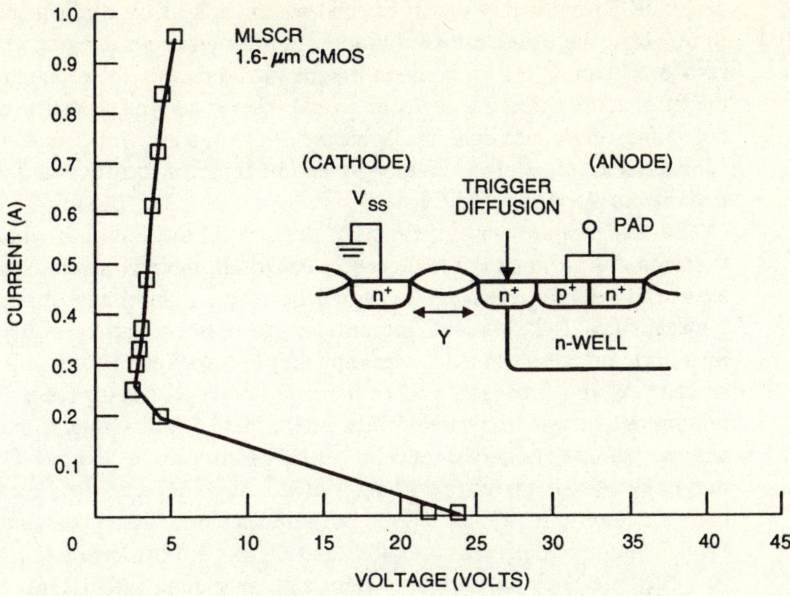

Figure 4.29   *I–V* characteristics for MLSCR with 1.6 μm CMOS process.

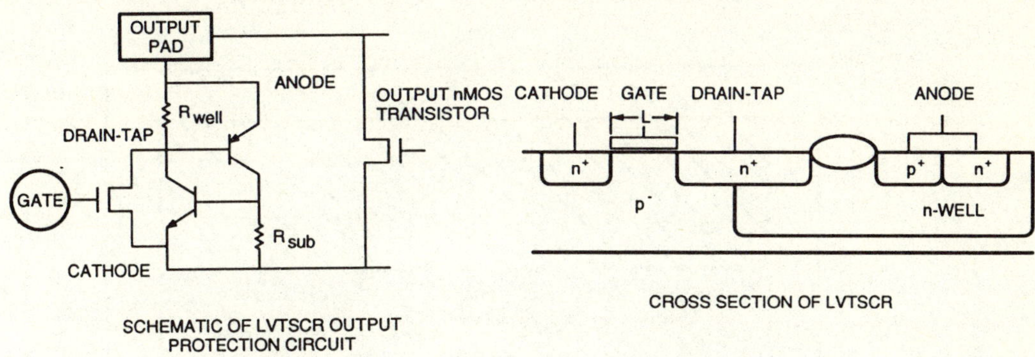

Figure 4.30   Cross-section of a low voltage trigger SCR (LVTSCR). The circuit equivalent is shown on the left.

solved either by making the channel length of the MOS device in the LVTSCR shorter than the output device or by placing an isolation resistor between the output device and the SCR protection device. The latter option has been successfully implemented in large submicron circuits [Carbajal92], but requires that the performance degradation of the output buffers caused by the addition of the resistor is compensated in the circuit design. An alternative is to reduce the trigger voltage of the SCR further by techniques similar to those used for the GCNMOS. By raising the gate voltage of the LVTSCR the voltage at which the SCR turns on can be significantly reduced [Diaz94]. This technique requires careful tuning to ensure that the trigger circuit does provide the correct triggering for the SCR.

The SCR is severely disadvantaged when used in floating substrate technologies. In this case the forward-biased diode is absent when stressing the pad negative with respect to ground. Thus, while in the forward direction the protection is provided by the SCR action, the *npn* needs to provide the protection in the reverse direction. To overcome this effect dual SCRs have been proposed which essentially have SCRs connected in parallel providing protection for both positive and negative polarity ESD stress voltages [Ker92].

The design and layout of the SCR devices follow a slightly different set of rules from the thin or thick oxide devices. First of all, there is no contact to drain spacing issue and usually all spacings should be at their minimum allowable values. The typical MLSCR layout and the critical layout spacings are noted in Figure 4.31. Spacing A determines the leakage. Spacing B is not very critical and can be collapsed as long as the process is non-silicided. For a silicided process, B is kept at a minimum to improve the gain of the lateral devices. Spacings C and C' are process-defined parameters and should be kept at a minimum. Spacing D will control the trigger since it is the channel length of the MOS device. For best trigger D is also kept at a minimum. Typical values of these spacings for a 1 μm technology are 4 μm for A, 1 μm for B, 0.5 μm for both C and C', and 1.6 μm for D. The MLSCR gate can be either metal or polysilicon. These similar spacings also apply to the LVTSCR, except the channel length can be reduced to 1 μm. Similar to the total protection using the thick oxide device, the SCR protection layout is shown in Figure 4.32 for a non-silicided process and in Figure 4.33 for a silicided process. Note that the series

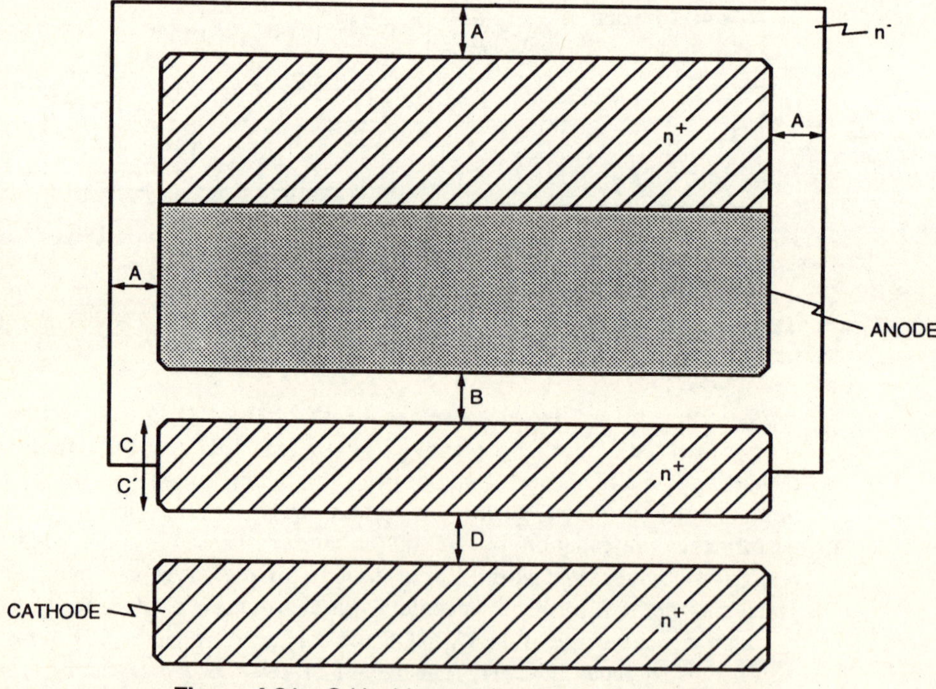

Figure 4.31 Critical layout parameters for an MLSCR.

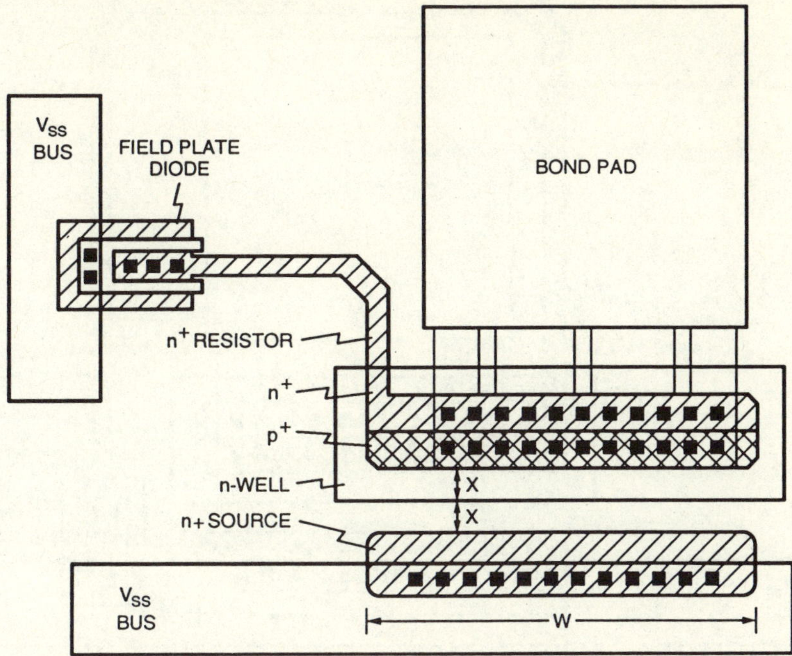

Figure 4.32 Layout example for input protection scheme using LSCR in a non-silicided technology.

resistor needs to be much longer for the silicided case due to the low sheet resistivity of the silicided diffusion. A different layout style for the nonsilicided protection scheme is described in Section 4.3.

In this chapter several different protection structures have been described giving design and layout details for each. However, their most effective protection performance is only determined when they act in a composite scheme as the primary devices. To achieve this, a design synthesis is needed. With this approach the most efficient composite protection scheme not only offers the best possible protection but also minimizes the input transit delay.

4.3 ESD PROTECTION DESIGN SYNTHESIS

In this section a synthesis of an advanced input protection scheme for applications in CMOS technologies is presented. The overall effectiveness of any input protection scheme is determined by the design of the constituent elements of the primary and secondary protection circuits. The SCR as a primary protection device was first introduced for bipolar technologies by Avery [Avery83]. For advanced CMOS technologies the lateral SCR (LSCR) has been shown to be effective as discussed in the last section. However, the design of the secondary protection to work in conjunction with this LSCR is not straightforward and can lead to failures [Rountree88][Duvvury89]. This is because the typical SCR trigger level tends to be high

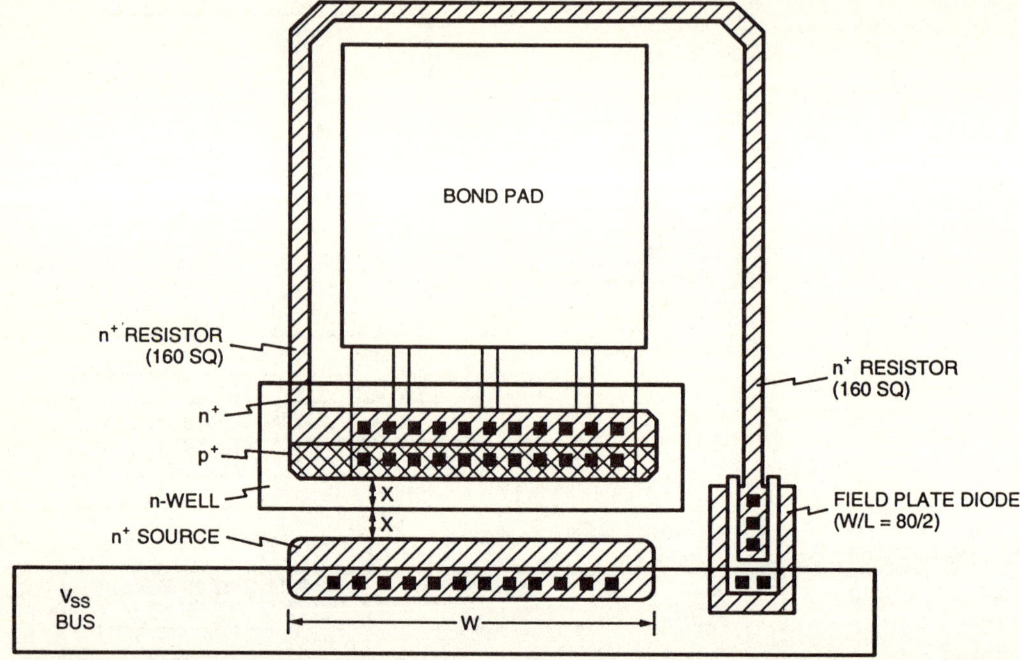

Figure 4.33 Layout example for input protection scheme using LSCR in a silicided technology.

(> 50 V) and the design issues involved for the secondary protection have not been clearly understood. In this section, the details of the secondary protection design are discussed. Moreover, an effective design with a combination of polysilicon resistor and SCR will also be discussed. This latter scheme can be an attractive option in analog circuit applications.

The approach taken here will be to analyze the individual protection elements of an overall input scheme, investigate how these work in conjunction with each other, and establish the ideal total combination. In all cases, pulse testing data is presented to understand the device response under ESD conditions. Failure analysis will be used wherever applicable to demonstrate the protection circuit functions.

We will briefly review the basic input protection scheme first. As seen earlier, for MOS technologies an input protection design consists of a field oxide device operating as a lateral *npn*. This device is combined with a grounded gate nMOS transistor (or FPD) through an isolation resistor to form the total input protection. The scheme is shown in Figure 4.34. During the initial ESD pulse, the pad voltage rises until the voltage across the FPD reaches the junction breakdown voltage. The lateral *npn* associated with the FPD is then turned on and as the current through the device increases, the voltage dropped across the resistor increases the pad voltage accordingly. When the pad voltage reaches the junction breakdown voltage of the field oxide device (FOD), the associated *npn* is triggered and the current is shunted through the FOD. The FOD turns on at about 30 V in a non-silicided 2 μm process.

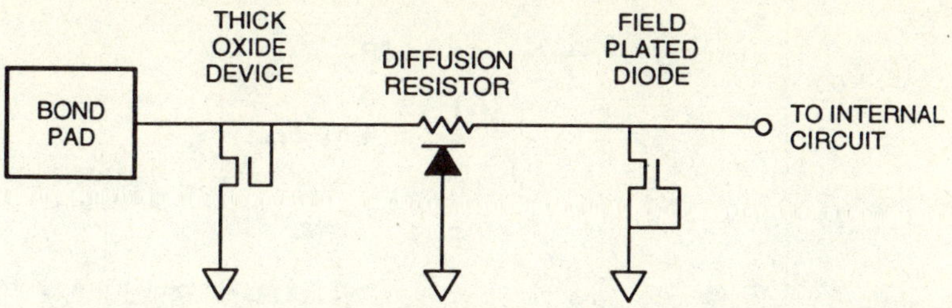

Figure 4.34   Input Protection Scheme with thick oxide transistor.

As discussed previously, the FOD has a higher intrinsic ESD capability than the FPD and, if the design of this device is optimized, it can yield more than 40 V/μm of width for the Human Body Model stress [Rountree85].

It was also observed that, in advanced processes, the thick field device performance can degrade considerably. If lightly doped drain (LDD) junctions are employed for source and drain, the protection per unit device width can degrade to 20 V/μm although it can be restored to about 30 V/μm with modifications to the source/drain diffusions [McPhee86]. The inclusion of silicides for the source and drain diffusions will reduce protection levels drastically down to 10 V/μm with little dependence on process or design parameters [McPhee86]. To overcome this, an SCR protection device can be an attractive option for CMOS processes [Rountree88]. However, an optimum choice of this device in conjunction with the isolation stage design is essential for an efficient input protection. These issues are discussed in the next two subsections. The results given here are mainly for a 1.6 μm CMOS process that does not employ LDD junctions. Nevertheless, the results are equally applicable to other CMOS and BiCMOS processes for VLSI circuits.

4.3.1 SCR primary protection

The trigger voltage of the LSCR, V_t, shown in Figure 4.27 is between 40 V and 70 V. As a result, if this device replaces the thick field device as the primary protection device in Figure 4.34, the secondary stage of the protection circuit, consisting of a diffusion resistor/FPD, must support the ESD stress until the pad voltage is high enough to trigger the SCR. If the secondary stage design is not optimal, it can lead to failure windows as reported by Duvvury [Duvvury89]. This problem can be eliminated if V_t is lowered.

We saw earlier that V_t can be substantially lowered by placing an n^+ diffusion at the n-well boundary, labeled *Trigger Diffusion* in the inset of Figure 4.35. The *I-V* curves of the MLSCR device are compared to the LSCR in Figure 4.35, also fabricated in a 1.2 μm CMOS LDD process. As expected, V_t is reduced from 50 V to 25 V for the MLSCR. Note that this value roughly corresponds to the *npn* breakdown of a thick field device for this process. When the *I–V* curves of both LSCR and MLSCR are compared in Figure 4.35, the on-resistance of the MLSCR is relatively larger. Considering that the anode–cathode space is the same for both

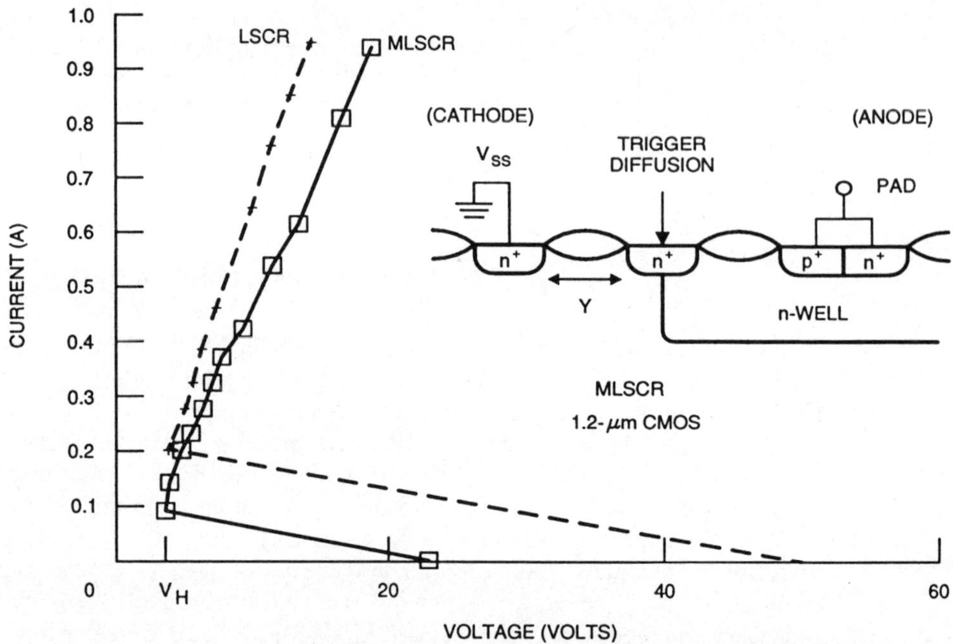

Figure 4.35 *I–V* characteristics for MLSCR (solid line) compared with *I–V* characteristics of LSCR without the trigger diffusion (dashed lines), for 1.2 μm CMOS process with LDD junctions.

cases, this behavior in the MLSCR could be due to the additional impedance in the device conduction path introduced by the n^+ at the well boundary. However, the ESD performance of both devices exceeds 6 kV with 100 μm of device width. An additional point for the MLSCR is that its trigger voltage is dominated by the n^+ avalanche threshold, whereas for the LSCR, the n-well to substrate avalanche breakdown voltage is very high. It is usual that punchthrough will occur before the avalanche breakdown voltage is reached and hence V_t is determined by the n-well overlap of anode, denoted as X in Figure 4.27. The variation of V_t with X for the LSCR was reported by Rountree *et al* [Rountree88].

4.3.2 Secondary protection devices

The secondary protection device needs to be able to carry some current before the primary protection device is triggered. For efficient protection circuit design, the devices used in the secondary protection have to be optimized. In this subsection we examine the use of a secondary protection scheme consisting of a resistor and a FPD. The two elements are considered separately and then the combined performance of the two is discussed.

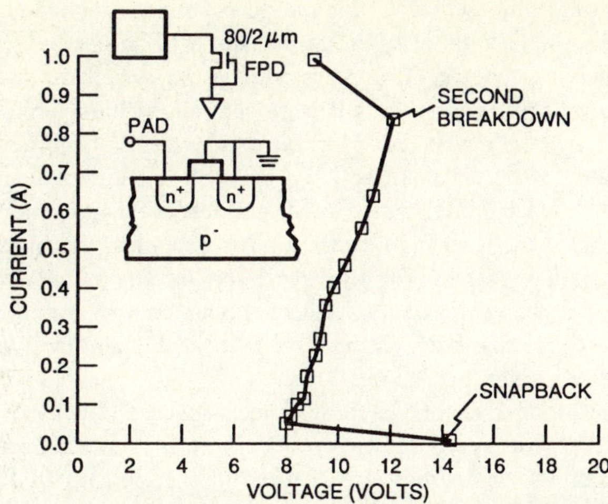

Figure 4.36   *I–V* characteristics for an 80/2 μm FPD device with 1.6 μm CMOS process.

Field plate diode

The *I–V* characteristics of an 80/2 μm grounded gate nMOS device are shown in Figure 4.36. The device goes into *npn* snapback after the drain avalanche break-down at about 14 V. At higher current pulses the device eventually enters the thermal second breakdown region which eventually leads to failure. The ESD failure thres-hold level has been correlated to the second breakdown trigger current level [Pol-green89][Amerasekera90]. Considering the 1.5 kΩ for the Human Body Model (HBM), the failure current level of 800 mA in the figure, which is obtained with the 150 ns wide constant current pulse (see Chapter 2), corresponds to about 1200 V HBM failure threshold. This translates to 15 V/μm for the 1.6 μm non-silicided process. In a 1 μm silicided process, this figure of merit can reduce to 4–5 V/μm [Polgreen89]. Referring back to Figure 4.36, the FPD can be effective in the input protection scheme as long as it is prevented from going into second breakdown. The design of the resistor and the primary SCR should take care of this criterion as discussed below.

Isolation Resistor

A diffusion resistor is commonly used for isolation stage protection. However, the diffusion resistor also acts as a parasitic diode to the substrate and may not support the voltage needed to trigger the primary device such as the SCR. That is, the effective value of this resistor can become only a fraction of the designed value. Hence, a large resistor is required to eventually build up the pad voltage for SCR trigger. Moreover, due to the resistor diode breakdown to the substrate, the contacts to the resistor easily get damaged. This situation can be improved by placing an *n*-

well around the resistor at the pad to suppress avalanche of the resistor diode. With this technique, only a minimum resistor will be needed for the overall protection.

A second type of diffused resistor is the *n*-well resistor [Carbajal92]. The main benefit of this resistor is its current saturation characteristic discussed in Chapter 3. Hence, a small resistance at low current levels can become very high as the current is increased, which is an attractive property to the circuit designer. However, the *n*-well resistor also has a negative resistance characteristic and will snap back to a low impedance mode at high voltages. The design of the resistor needs to comprehend the snapback problem for effective behavior. The main advantage in using the *n*-well resistor is the current saturating phenomenon which is determined by the width of the well. The snapback voltage is defined by the length of the resistor. Some numbers have been given in Chapter 3.

Polysilicon resistors have also been successfully used as part of the protection circuits [Duvvury83]. However, since these elements are encapsulated in low thermal conductivity oxide they are thermally isolated and the power dissipation causes damage at relatively low ESD levels. As reported by Fukuda and Kato [Fukuda88], they can be used to improve the Machine Model performance.

4.3.3 Protection scheme

The constituent protection elements can be combined to form the total input protection circuit. The secondary protection is first examined before combining the SCR to form the full protection scheme.

The *I–V* characteristic when the avalanche-suppressed diffusion resistor is combined with the FPD, is shown in Figure 4.37. Immediately after the nMOS of the FPD breaks down, the current through the resistor causes an increase of the voltage at the pad. For the secondary protection layout (see Figure 4.42 later), a parasitic thick field device (formed with n^+ diffusion of the resistor, *p*-substrate, and

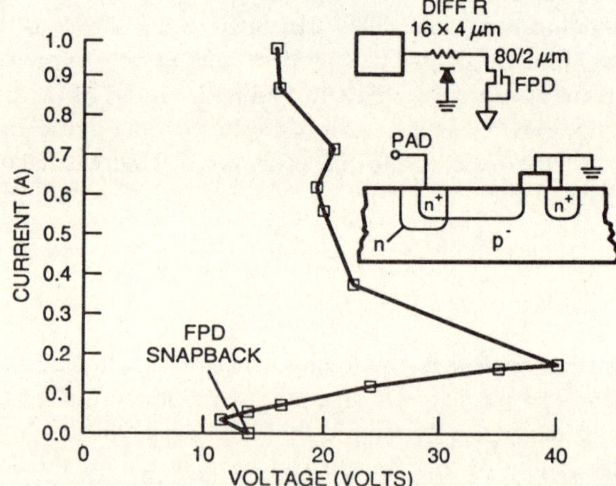

Figure 4.37 *I–V* characteristics for secondary stage protection with 16 × 4 μm diffusion resistor and 80/2 μm FPD device. The process is 1.6 μm CMOS.

n^+ connected to V_{ss} of the FPD) turns on when the n^+ to substrate avalanche breakdown voltage is reached. This is one of the disadvantages of using a diffusion resistor as will be discussed later. In agreement with this assumption the clamping voltage approaches 15 V, typical for the snapback for a thick field *npn*.

During ESD stress, after the *npn* snapback occurs at the FPD, the voltage at the high end of the resistor near the pad increases until a failure occurs. If the FPD were removed, all the current would go through the resistor/diode to the substrate. With simple assumptions about the substrate resistance and diode avalanche, it can be shown that in this case the power dissipation would be higher at the pad side of the resistor. Thus the protection level should be higher for the resistor plus FPD combination than with the resistor/diode alone, if the limiting failure mechanism is the resistor damage. First of all, the HBM failure threshold distributions are shown in Table 4.1 under the column labeled 'V_f (without SCR)'. Comparing case 2 with case 3, the general increase in the protection level with inclusion of the FPD device is quite clear. It is also interesting to note from case 1 that the FPD by itself can offer some protection which is enhanced by adding the resistor/diode combination. Note that case 3 in general seems to be a cumulative effect of case 1 and case 2. From case 4 in the table, it is also seen that the diode/resistor has a limited effect on improving the failure level since increasing the resistor size does not further enhance it.

These results become clearer when the failure modes are examined. In Figure 4.38(A) the failure mode for the resistor without the FPD is shown. The damage is seen at the end contacts indicating the suppression of avalanche due to the *n*-well. The region outside the *n*-well avalanches and eventually electrical failure occurs when the damage region reaches the substrate. Now if the FPD is added, the damage is shown in Figure 4.38(B). Here, for an electrical failure, the damage occurs both at point *A* in the resistor (similar to damage in Figure 4.38(A)) and at point *B* in the FPD. The damage at point *B* is simply the common gate-drain short. What is more interesting is that the resistor damage is confined in the resistor body and cannot be electrically detected until the short occurs at the FPD. This is an example of how electrical failure detection cannot always reveal the pre-threshold failures. Incidentally, the failure voltage distribution for case 3 corresponds to only resistor damage at the low end and both resistor and FPD damage at the high end. That is, in

Table 4.1 Input protection performance with different choices of secondary protection design.

CASE NO.	R (DIFFUSION) (OHMS)	W/L OF FPD (μm)	V_f (WITHOUT SCR) (kV)	V_f (WITH SCR) (kV)
1	–	80/2	0.9 - 1.2	0.9
2	150	–	1.2 - 2.0	>6
3	150	80/2	1.9 - 3.1	>6
4	240	80/2	2.0 - 2.5	>6

CASE NO.	R (POLYSILICON) (OHMS)	W/L OF FPD (μm)	V_f (WITHOUT SCR) (kV)	V_f (WITH SCR) (kV)
5	85	80/2	0.9 - 1.0	>6
6	150	80/2	0.8 - 1.0	>6
7	220	80/2	0.7 - 1.0	>6

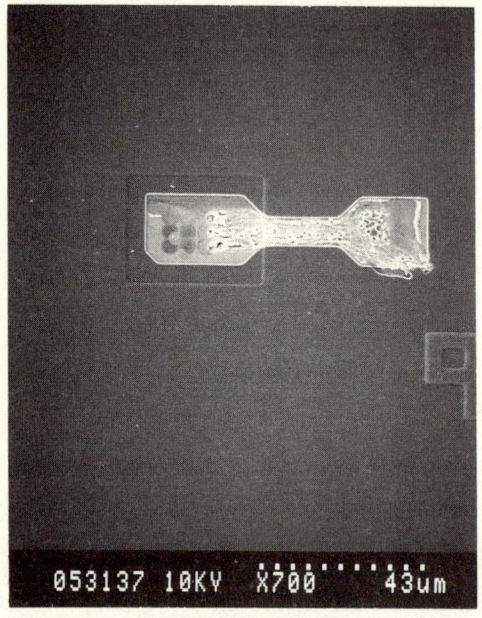

(A) WITHOUT FPD

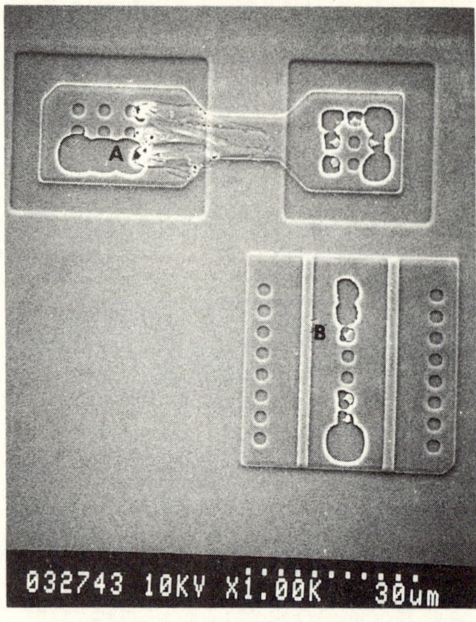

(B) WITH FPD

Figure 4.38 Failure sites for secondary protection with diffusion resistor. The damage shown in (A) is for failure in the diffusion resistor without the FPD at a stress level of 2 kV. In (B), the damage locations with the FPD included are shown at a stress level of 3.1 kV. Note that in (B) the damage appears in both the resistor (**A**) and in the FPD (**B**).

some cases as noted above, if the damage is confined to the resistor only, failure level goes up to 3 kV before the electrical damage is detected.

Instead of the diffusion resistor, a polysilicon resistor may also be used. The $I–V$ curve for a combination of polysilicon resistor and FPD is shown in Figure 4.39. After the FPD snapback the current through the resistor causes an increase in voltage at the pad. Note that the pad voltage build-up is continuous without any breakdown. This is expected since no parasitic devices are present and the resistor supports the full voltage. A more detailed analysis of different-sized polysilicon resistors is shown in Figure 4.40. At the higher current levels the resistor seems to heat up causing an increase in the resistance, where the onset of heating is indicated by arrows. Note that the heating effect seems to be minimum for the widest resistor. In contrast to this, the heating effect is not observed for the diffusion resistors as shown in Figure 4.41 where much of the current is due to avalanche breakdown of the diode to substrate. Although there was no evidence that heating in the polysilicon resistors has any adverse effect on its performance, for practical applications the safe region of operation should be kept below the onset of heating. Hence, the ESD protection design can be achieved by keeping the polysilicon resistor operation in the region of low power dissipation.

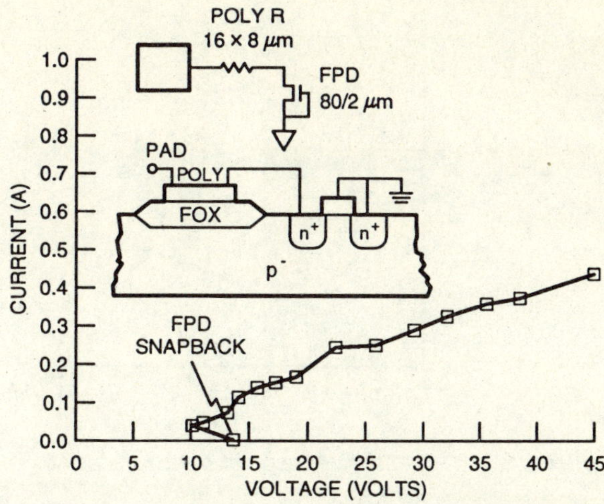

Figure 4.39 *I–V* characteristics for a secondary stage protection with 16 × 8 μm polysilicon resistor and 80/2 μm FPD device. The process is 1.6 μm CMOS.

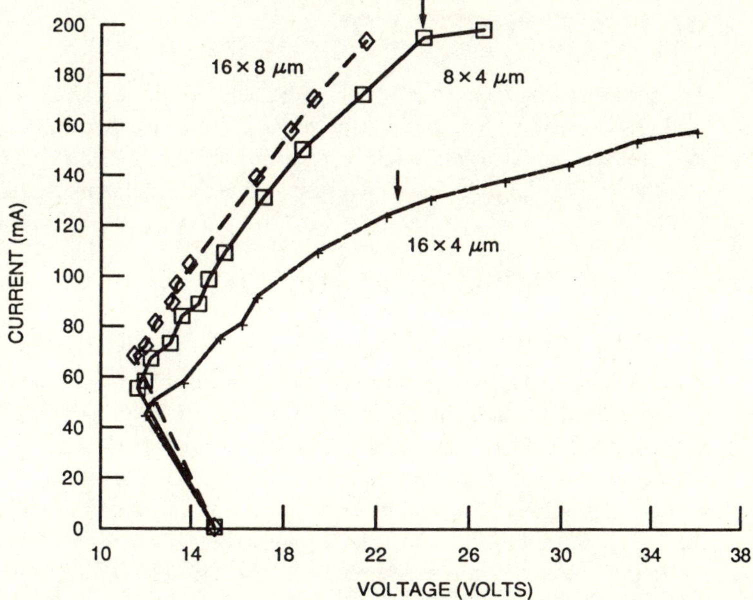

Figure 4.40 *I–V* characteristics of polysilicon resistors combined with an FPD. The arrows point where heating effects seem to begin.

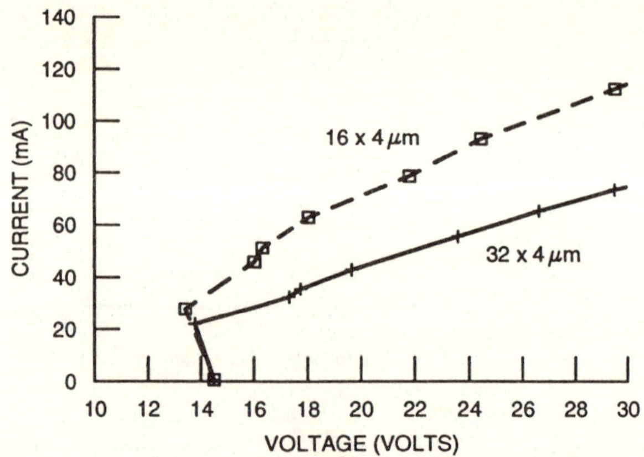

Figure 4.41 I–V characteristics of diffusion resistors combined with an FPD.

The HBM failure voltages for polysilicon resistors are also shown in Table 4.1 (see 'V_f (without SCR)'). Although there is some scatter in the data, it is clear that the minimum failure level increases for the smaller resistor. The differences are not distinct because the failure level of the FPD is very close (see case 1). But it is clear that, unlike for the diffusion resistor, adding the FPD does not improve its failure level. This shows that either the polysilicon resistor or the FPD fails around the same point.

4.4 TOTAL INPUT PROTECTION

The protection elements have been characterized in the previous sections with pulse testing. The ideal combinations of these for an effective input protection are discussed here. Both the diffusion resistor and the polysilicon resistors are considered separately. For the polysilicon resistors the reliability aspects are also examined.

4.4.1 Inputs with diffusion resistor

The design with a diffusion resistor becomes effective when the n-well is placed to suppress avalanche. A layout of such a circuit is shown in Figure 4.42. Note that the layout is arbitrary and can be changed to minimize the parasitic device effects. The matrix layout of contacts achieves the optimum performance where the current density through each contact is reduced. The I–V curves for the MLSCR for this particular process of the total protection circuit is shown in Figure 4.29. Compared to the MLSCR of Figure 4.35, the trigger point is about the same but the holding voltage is lower. It should be noted that the *Trigger Diffusion* in this application is butted against the p^+ anode, but this modification has not been found to have any impact on the SCR operation itself. The overall I–V curves of the protection circuit

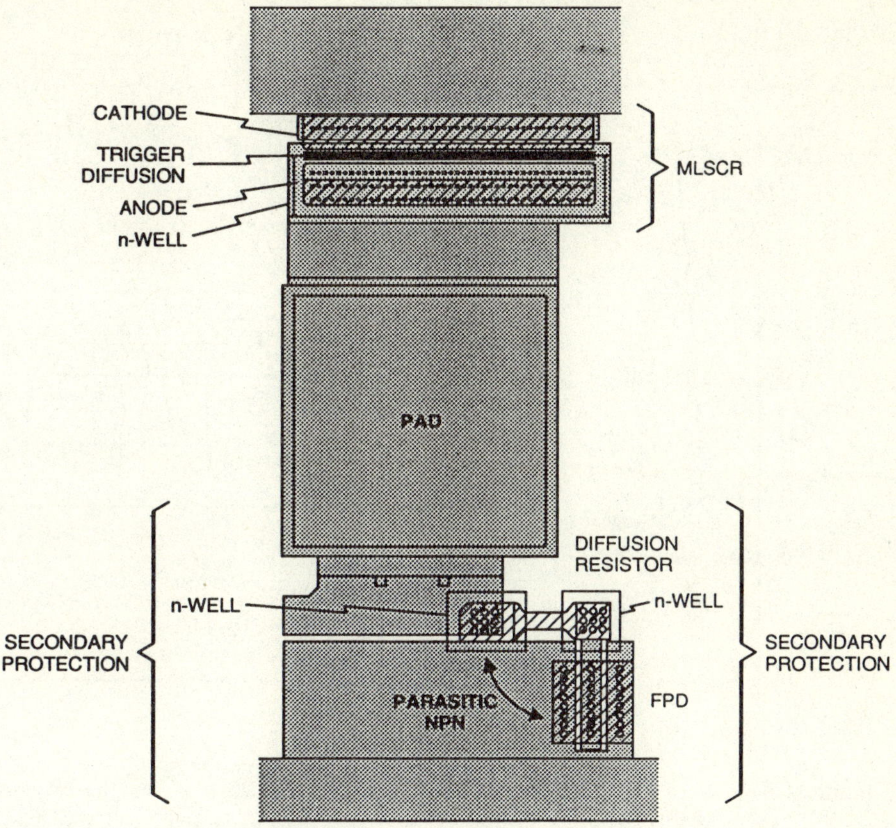

Figure 4.42   Layout of a total protection with diffusion resistor and MLSCR. Note the parasitic *npn* device.

with a 150 Ω diffusion resistor are shown in Figure 4.43. When the current level reaches 100 mA the voltage at the high end of the resistor approaches 25 V (see Figure 4.42), which is equal to a 15 V drop across the resistor plus the FPD snapback voltage of 8 V. As expected, this triggers the MLSCR device and the current level abruptly increases by 200 mA.

Design of the secondary element and the resistor essentially follows the following methodology. Assuming that a FPD is used as the secondary element, the maximum current through the FPD must be determined by characterizing these elements for a given process. If the maximum current is I_{t2} in mA/μm where I_{t2} is the second breakdown trigger current (Chapter 6), then the maximum allowable current should be guardbanded to about $0.75 I_{t2}$. The snapback holding voltage of the FPD, V_{sp}, is also determined in the characterization, as is the trigger voltage of the primary device, V_{trig}. The value of the resistor, R, to be used is then determined by

$$R \times W \geq \frac{V_{trig} - V_{sp}}{0.75 I_{t2}} \qquad (4.1)$$

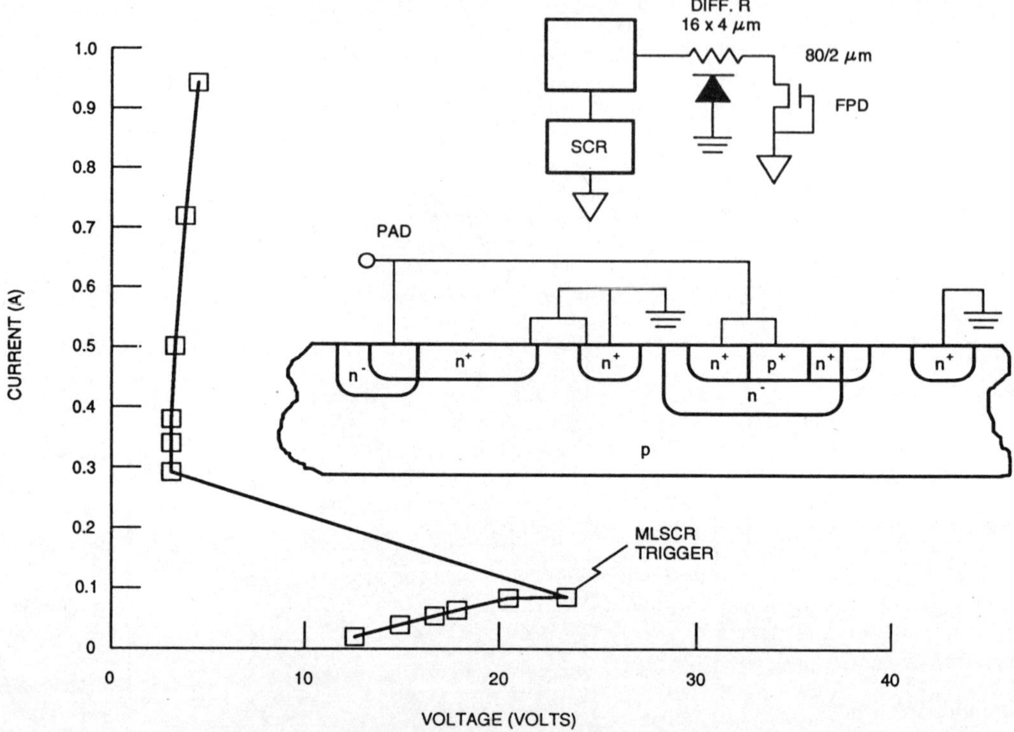

Figure 4.43　$I–V$ characteristics for total protection with 16×4 μm diffusion resistor, 80/2 μm FPD, and 150 μm wide MLSCR. The process is 1.6 μm CMOS.

The HBM ESD performance by combining with MLSCR is summarized in Table 4.1 under 'V_f (with SCR)'. Considering Case 1, it is obvious that the FPD by itself cannot trigger the SCR device since its breakdown voltage is lower. Thus failures occur at low levels corresponding to the FPD protection level. In Case 2, the diffusion resistor (with avalanche suppression) alone in parallel with the SCR cannot obviously have an impact on the SCR function. However, for protection of the input gate oxide, an FPD device is also needed. Thus only for Case 3 when both the FPD and the resistor are combined, full effective SCR protection for inputs is achieved.

For maximum circuit speed at the input, the isolation resistor needs to be as small as possible. With a regular LSCR of 50 V trigger, this cannot be easily achieved for an n-diffusion resistor. Even with avalanche suppression the full resistor value is not realized and failures occur unless the resistor is made more than 100 Ω. However, when the avalanche-suppressed resistor is used in conjunction with an MLSCR, a minimum necessary value for the resistor can be used. n-well resistors will also have limitations in the maximum voltage across the resistor before snapback occurs.

4.4.2 Inputs with polysilicon resistor

The input combination with the polysilicon resistor is shown in Figure 4.44 with the measured $I-V$ curves in Figure 4.45. As before, after the FPD snaps back the pad voltage builds up through the $I \times R$ drop across the resistor to trigger the MLSCR. As long as the SCR trigger occurs below the failure current, the protection scheme would be effective. The curves of Figure 4.45 were measured for the total protection employing the 16×8 μm polysilicon resistor with its effective resistance of 85 Ω as shown in Figure 4.40. The MLSCR is observed to trigger at a low current level of 150 mA (or approximately 250 V HBM). Equation 4.1 above is valid for polysilicon resistors as well. An optimum value for the resistor, considering the 15 V drop needed, would be 60 Ω. Values smaller than this would not allow sufficient pad voltage build-up and values larger than 100 Ω would increase the chances for some latent damage in the secondary protection. It is interesting to note in Figure 4.45 that, following the MLSCR trigger, the current level increases by about 330 mA, which is more than for the diffusion resistor case in Figure 4.43. This indicates that

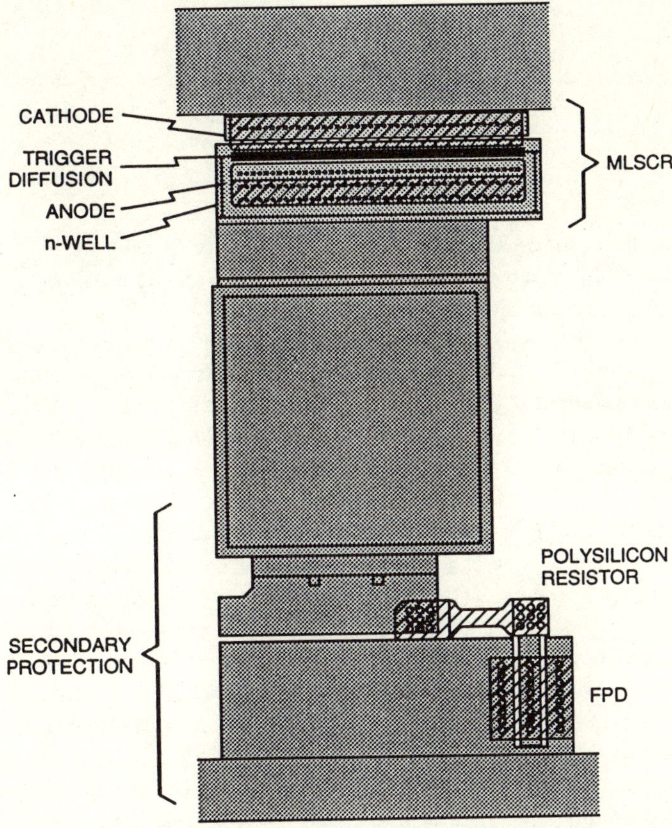

Figure 4.44 Layout of a total protection with polysilicon resistor and MLSCR.

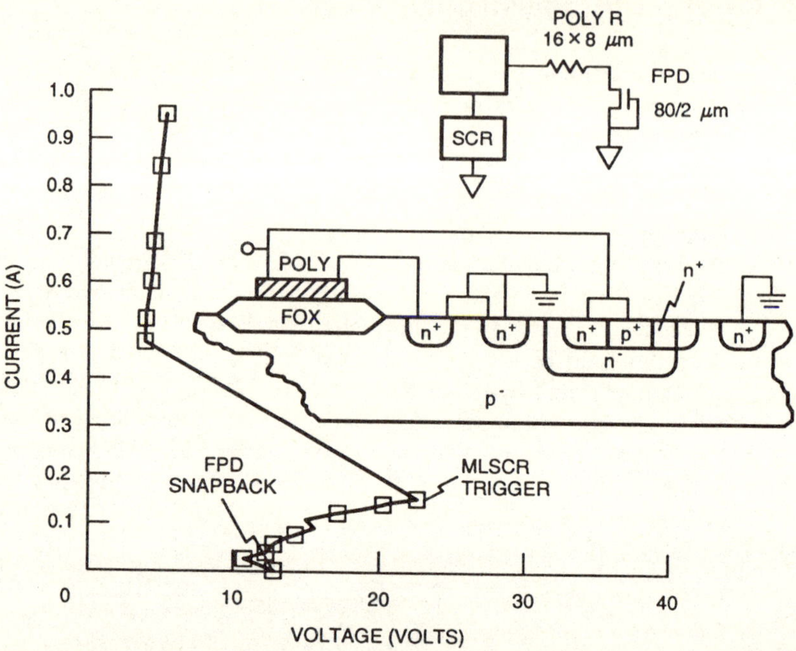

Figure 4.45 *I–V* characteristics for total protection with 16 × 8 μm polysilicon resistor, 80/2 μm FPD, and 150 μm wide MLSCR. The process is 1.6 μm CMOS.

after the MLSCR triggers, most of the current flows through this device, removing all of the stress from the polysilicon resistor. This is important for effective and consistent performance of this scheme.

The HBM ESD performance with the polysilicon resistor is summarized in Table 4.1 under 'V_f (with SCR)'. The protection for the polysilicon resistors in Cases 5–7 all worked well. This is not surprising since the needed voltage drop of 13 V can be achieved below their failure thresholds without the SCR device. A minimum value between 60 and 100 Ω can be easily selected for minimum RC delay effects at the input. A smaller resistance can be used but the width will have to be made greater than 8 μm. In this case, the trade-off is between the area and the input delay has to be evaluated. An even lower resistor value can be used if an LVTSCR is combined instead of the MLSCR. This is because the LVTSCR can have a low trigger voltage in the range of 12–16 V depending on the process. This protection design with the MLSCR was also evaluated for negative stress with no failures for greater than 6 kV stress. For this polarity stress, apparently the forward diode in the MLSCR device clamps the voltage and does not allow any significant conduction of the diode from the FPD device. Testing with the Machine Model has shown that this protection scheme with polysilicon resistor/MLSCR is very effective for this stress also, with a failure threshold greater than 1000 V. This is well in excess of the usually required 200 V for the Machine Model.

4.4.3 Polysilicon resistor reliability

The reliability of the polysilicon resistor in a protection circuit would be of great concern since damage to it may not be electrically detected. In order to evaluate this, the secondary protection scheme with a polysilicon resistor was stressed at different HBM levels. The stressing was done with 10 pulses in each case to give confidence to the results. The SEM photographs of the stressed devices and the stress levels are shown in Figure 4.46. It is interesting to note that for both 200 V and 400 V stress levels there is no physical damage to the polysilicon resistor. At the 600 V level, the first indication of heating is seen at the tapered portion of the resistor. It is possible to improve the layout to reduce this effect. Nevertheless, at higher stress levels the resistor is eventually damaged. What is more interesting is that at 800 and 900 V levels there is obviously severe damage to the resistor that was not electrically detectable since a continuous polysilicon filament is present. This is another demonstration of the deceptive nature of pure electrical analysis. It is interesting to note here that for the 900 V stress level some contact damage is also present, indicating

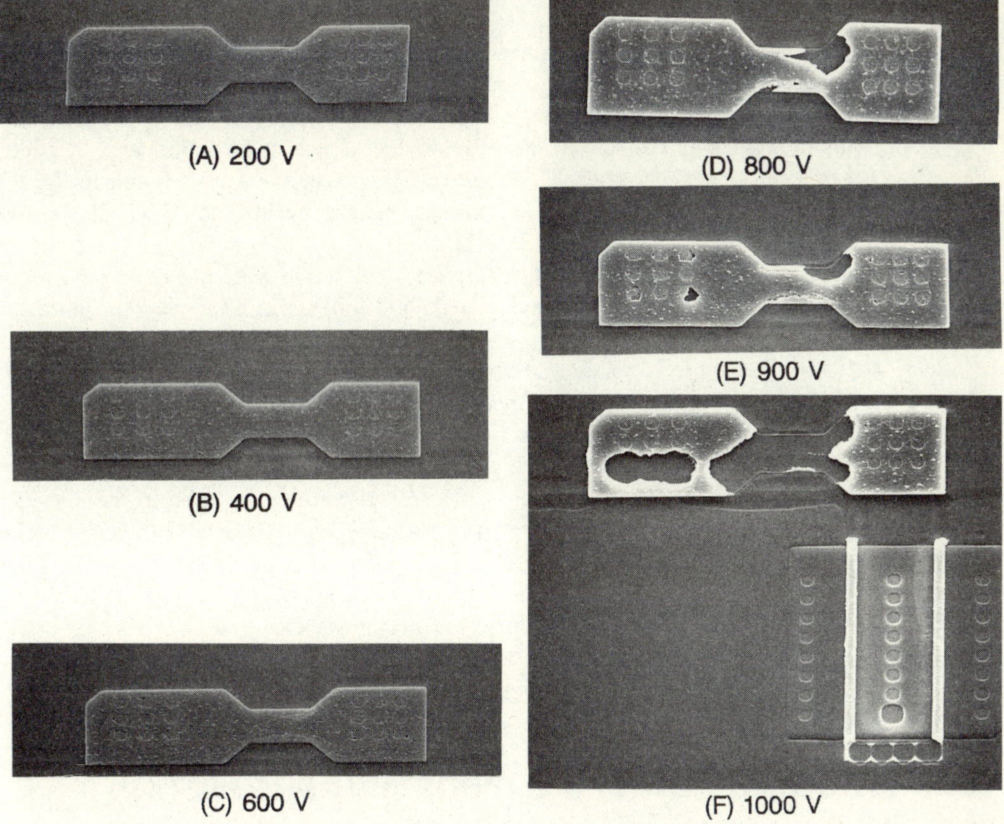

Figure 4.46 Detailed failure analysis of secondary protection with 8×4 μm polysilicon resistor, showing progressive ESD damage. The stress conditions for HBM are: (A) 200 V, (B) 400 V, (C) 600 V, (D) 800 V, (E) 900 V, and (F) 1000 V. The initial damage for 600 V stress is indicated by the arrow.

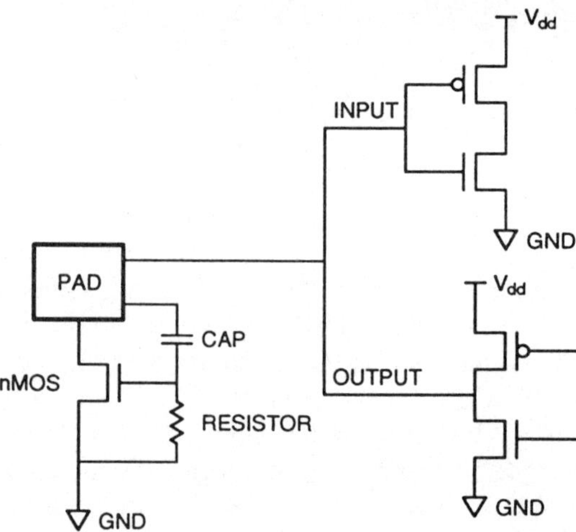

Figure 4.47 An I/O protection scheme using the GCNMOS.

that the current density through each contact exceeds the failure level. This can also be improved by increasing the number of contacts. At the 1000 V stress the polysilicon is completely blown giving rise to an open failure. In this experiment in all cases there was no damage to the FPD device. However, in some cases at 1000 V stress level, damage to the polysilicon resistor at the taper corner and to the FPD occurred simultaneously. This is because the failure level of the FPD is also close to this stress level (see Table 4.1, Case 1).

From the analysis of Figure 4.46, the total protection with an SCR would require that the polysilicon resistor survive at least 400 V of stress. Referring to Figure 4.45, the SCR triggers at 150 mA or approximately 250 V for the HBM. This clearly indicates that there is a safe margin between the SCR trigger current (at $<$ 250 V HBM) and the current at which thermal damage to the resistor occurs (at $\approx$600 V HBM). The margin could be higher considering that the analysis in Figure 4.47 was done for an 8 $\times$ 4 μm resistor which has a higher tendency to heat up (see, for example, Figure 4.42) than the 16 $\times$ 8 μm resistor actually used in the protection design. Moreover, the layout of the 16 $\times$ 8 μm resistor did not have the taper regions to further reduce the heating effects. Finally, the total protection circuit of Figure 4.45 was stressed at $\pm$ 6 kV for 20 times to note the effectiveness of the polysilicon resistor. As expected, there was no physical damage to the resistor after this stress. This clearly shows that a polysilicon resistor with an SCR device can be used with confidence, if the proper design synthesis is used.

4.5 INPUT/OUTPUT BUFFER LAYOUT AND PROTECTION

CMOS Input/Output buffers require ESD protection for both gate oxides as well as the diffusions connected directly to the pad. In general, if the output buffer diffusions are well protected, the input buffer is also well protected. The nMOS pull-down transistor in the output buffer is the weakest link in CMOS input/output buffers and the ESD protection effort needs to focus on these devices.

The layout of the nMOS output transistor should follow all the rules for the nMOS transistor discussed in Section 4.2.2. In addition, the protection levels are increased if dedicated protection circuits providing parallel shunt paths for the ESD current are included at the input/output pads as shown in Figure 4.47. Here the use of the GCNMOS as a protection device for input/output buffers is shown. In all cases the gate transient is designed to be about 10 ns. For output applications as a dummy device, a capacitor can be added between the gate and drain to ensure that the coupling on the protection device gate is always relatively higher than that for the ouput buffer device.

Instead of a GCNMOS, protection schemes using dual-diodes [Voldman92] [Dabral93][Dabral94][Voldman94] and SCRs [Chatterjee91A][Carbajal92][Diaz94] have also been reported. They all require that the nMOS transistors are laid out for uniform current flow and maximum power dissipation. In the case that full nMOS output buffers are used, both the pull-down transistor and the pull-up transistor must follow the ESD layout guidelines. These include requirements that the minimum width of an nMOS in the output buffer is 200 μm although this may be more for more sensitive processes.

The use of an isolation resistor between the primary protection circuit and the output buffers is not a desirable option from the point of view of normal circuit operation. It also means that the output buffers need to be increased in size to accommodate the performance degradation due to the isolation resistor. However, when it is possible to include isolation resistors, they have been shown to provide very high ESD performance levels for output buffers with dedicated primary ESD protection circuits [Carbajal92].

While the layout of the n-channel device is very critical for ESD, the p-channel device layout can also play an important role in the case of CMOS output buffers.

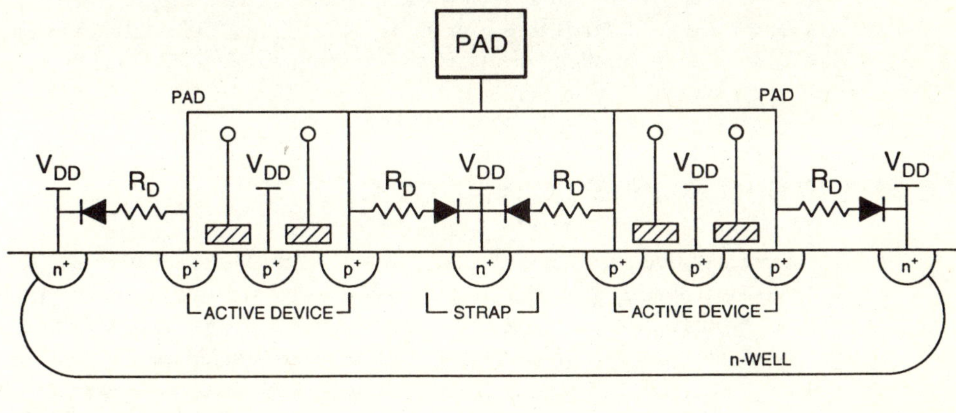

Figure 4.48   Cross-section of pMOS transistor with improved lateral diode to V_{dd} by placing additional n^+ straps.

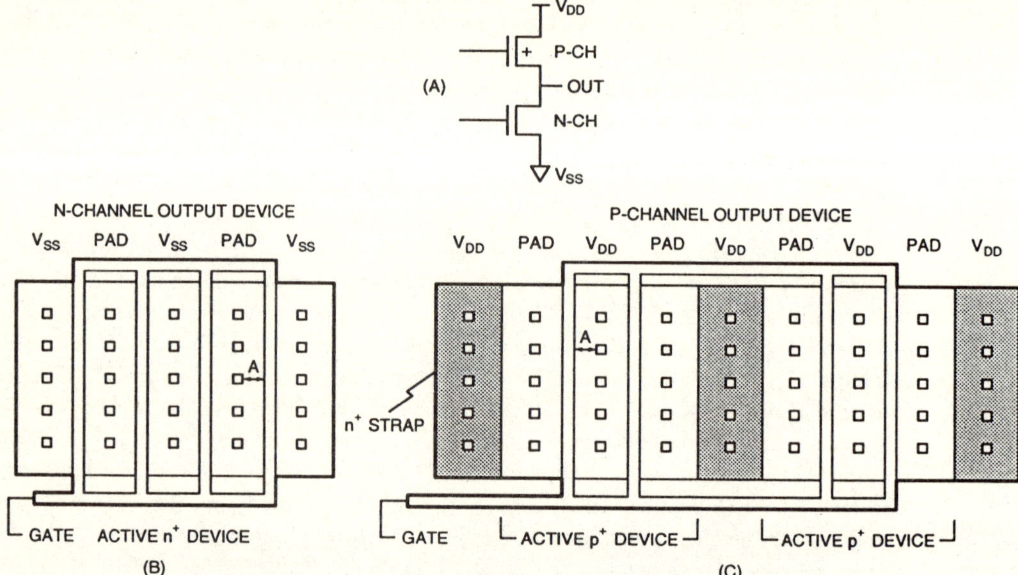

Figure 4.49 Layout of pMOS transistor with improved lateral diode to V_{dd} by placing additional n^+ straps.

With the p-channel present, the initial ESD current goes through the intrinsic pn diode formed between the p^+ diffusion and the n-well contacted to the power supply bus, V_{dd}. The power supply bus is connected to the entire n-well of the chip and forms a capacitance which can be $\gg 1$ nF in some large chips. The ESD current charges up the V_{dd} capacitance thus relieving the initial stress on the primary protection device which results in an improvement to the overall ESD levels [Duvvury87]. One way to lay out the pMOS device for this purpose is to add additional lateral diode paths as shown in Figure 4.48. This improvement consists of adding n^+ diffusion straps to V_{dd} in the layout of the pMOS transistor as shown in Figure 4.49. This technique has been shown to improve the ESD performance of a protection circuit [Duvvury88].

4.6 SELECTING A PROTECTION CIRCUIT

In the preceding sections the protection design with an MLSCR device was analyzed. It was shown that once the primary protection is chosen the secondary protection design becomes critical. A scheme with a diffusion resistor can be effective as long as avalanche suppression is employed using an n-well around the n^+ diffusion. Even without n-well the protection works well but requires a larger resistance value for safe design. The higher resistance will have an impact in circuit applications that demand high speed. With the n-well suppression in place the resistance can be much lower for the same ESD performance. Protections with the MLSCR and diffusion resistor, as described here and shown in Figure 4.42, should use a 100 Ω resistor and an 80 μm wide FPD. The channel length can be minimal, although 20% greater than the minimum channel length is recommended for reducing possible leakage effects at the inputs. The resistor width should be a minimum of 4 μm with the *n*-well overlap of diffusion at 6 μm.

The input protection can also be effective with the use of a polysilicon resistor in conjunction with an MLSCR. There are several advantages with this approach. First, the parasitic capacitance is significantly reduced to minimize the RC delay. Second, polysilicon resistors are not prone to the voltage coefficient like the diffusion resistors are. For example, in voltage divider designs with a series of diffusion resistors the depletion regions will vary giving rise to non-linear effects. But this is not the case for polysilicon resistors. In analog circuit applications a polysilicon resistor is preferred because of its better leakage performance at high temperatures. Finally, protection designs with a polysilicon resistor will have the additional advantage of being immune to parasitic device interactions or *proximity effects* as reported for a DRAM protection design [LeBlanc91]. In contrast to these advantages, a polysilicon resistor is highly susceptible to heat damage since it is encapsulated with no available heat sink from the substrate. Thus, there is always a possibility that any latent damage can go undetected. But as was shown in this section, careful analysis and design can eliminate such phenomena. The layout implementation should be as in Figure 4.44 but with the taper regions removed. In silicided processes a 4 μm wide polysilicon resistor should be adequate but no data have been reported to establish a definite recommendation.

4.7 BIPOLAR AND BiCMOS PROTECTION CIRCUITS

4.7.1 Introduction

The main protection circuit issues for bipolar processes are very similar to those described above for CMOS processes. The requirements of the protection circuit in terms of the trigger voltage, high current behavior, capacitive and resistive loading, and area constraints must all be taken into consideration during the design phase. BiCMOS processes which combine both bipolar and CMOS elements on the same chip can have benefits in terms of the available protection elements as well as disadvantages because of the complexity of the circuits to be protected. The present generation of bipolar and BiCMOS circuits are targeted at high speed applications such as telecom. These applications present additional constraints on the design of ESD protection circuits because of their inability to tolerate any additional capacitance at the input/output pads or resistance in the output transistors themselves. In this section we will briefly look at those protection circuit design issues specifically related to bipolar and BiCMOS circuits.

4.7.2 Protection circuit strategies

Large bipolar transistors usually have very good intrinsic ESD protection levels since they are designed to operate as vertical *npn* devices rather than the lateral parasitic *npn* transistors available in MOS processes. Typical ESD capabilities for an advanced bipolar *npn* transistor are about 30 V/μm compared to <15 V/μm for the lateral *npn* transistor in a CMOS process. These large *npn* transistors can be made self-protecting provided the basic ESD layout rules for uniform current flow and peak electric field reduction are followed. Some of these rules have already been discussed in Section 4.2.2 for nMOS transistors, and in this section we will discuss specific bipolar-related issues.

BiCMOS processes can have CMOS output buffers which are much weaker ESD elements than *npn* transistors [Amerasekera92][Tandan94]. In advanced BiCMOS processes, therefore, outputs employing nMOS transistors will be the limiting factor in good ESD performance. The focus of BiCMOS ESD protection circuit design should be to protect the nMOS transistors used in the input or output buffers. The available protection circuit elements are the vertical *npn* transistor as well as bipolar SCR structures. The lateral CMOS SCR is made ineffective by the low resistance buried n^+ diffusion at the bottom of the *n*-well used for the *npn* collector. It has been observed that the bipolar SCR does not gain any ESD performance over the *npn* transistor [Amerasekera92]. The *npn* has the added advantage that it is easier to implement directly into a circuit model such as SPICE for evaluation of both circuit performance as well as ESD performance [Chatterjee91B].

In order to use the *npn* transistor as a primary protection element for nMOS transistors in the output buffers, it must be ensured that the *npn* triggers before the nMOS transistor that is being protected. It must also have a lower snapback holding voltage and on-resistance than the nMOS. Lowering the snapback trigger voltage is achieved using active elements as triggers as described in [Chatterjee91B][Amerasekera92]. In the event that the *npn* transistor characteristics do not meet the requirements for protecting the nMOS transistor, even with a trigger circuit, nMOS protection design needs to be implemented or the nMOS needs to be made self-protecting as described in previous sections.

4.7.3 Bipolar/BiCMOS output protection

The triggering and operation of a vertical *npn* transistor during an ESD event has been described in Chapter 3. It was pointed out that by forward-biasing the emitter-base junction of the *npn* the voltage required to fully turn on the *npn*, V_{t1}, can be reduced. The schematic circuit shown in Figure 4.50 shows a vertical *npn* transistor T1 with its collector connected to the pad. T2 is a generic trigger element which turns on during an ESD event and allows current to flow though the resistor R. T3 is the nMOS device being protected. When the emitter-base voltage, V_{be}, is sufficiently high that the emitter current can substantially contribute to the avalanche process at the collector-base junction, the *npn* goes into a low impedance mode shunting the ESD current (Chapter 3). A rough value of $V_{be} \approx 0.8$ V will lower V_{t1} to significantly less than the trigger voltage of the nMOS. Based on this value of V_{be}, if T3 carries 1 mA then R needs to be about 1 kΩ.

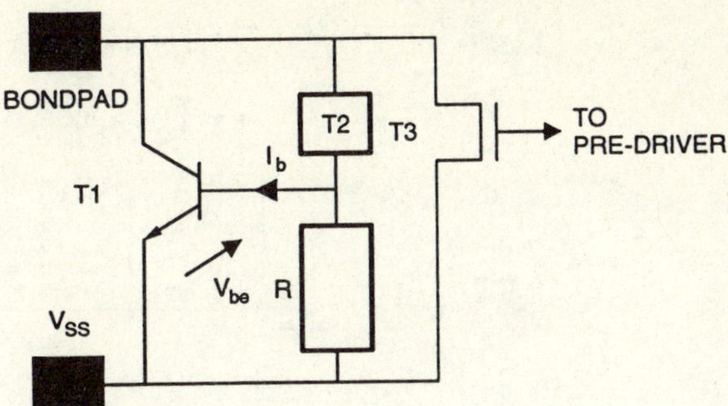

Figure 4.50   Bipolar/BiCMOS output protection scheme using active triggering [Chatterjee91B].

T2 can be an nMOS transistor, a reverse-biased diode or a lateral *pnp* transistor [Chatterjee91B][Amerasekera92][Corsi93]. A low breakdown voltage diode can be easily obtained in a bipolar process by using the emitter-base breakdown voltage of the *npn* which is typically between 5 V and 7 V. ESD levels of >4 kV have been demonstrated in a submicron BiCMOS process using this technique [Chatterjee91B].

In a bipolar output buffer, T1 can be the *npn* pull-down in the output buffer itself. Assuming a conservative ESD level of 20 V/μm for the *npn*, a minimum emitter length would be 200 μm for a 4 kV performance. If an *npn* is also connected between the pad and V_{dd} the ESD levels will be greatly increased. The ESD performance when the emitter (collector/base grounded) is stressed has not been found to be much less than that obtained with stress applied to the collector (emitter/base grounded). Conservatively, the ESD threshold for the emitter stress is $\sim$80% of that for the collector stress.

When Emitter-Coupled-Logic (ECL) buffers are used the power supply connections are different, but similar designs can be used with adjustments made for the different power supplies (i.e. $V_{cc} = 0$ V, $V_{ss} < 0$ V).

4.7.4 Bipolar/BiCMOS input protection

Figure 4.51 shows a schematic bipolar protection circuit for protecting a bipolar or MOS input buffer. The circuit design follows the design synthesis described for input protection in Section 4.3 and consists of a primary protection and a secondary protection separated by an isolating resistor. B1 is the primary protection using the *npn* transistor. B2 and B3 provide secondary protection for clamping the voltage at the input. These are effective for bipolar inputs, but MOS inputs require an FPD clamp N1 to protect against gate oxide breakdown. B3 provides a low voltage clamp between the pad and V_{cc}. R1 is the isolation resistor. These protection circuits have been shown to be capable of ESD protection levels > 4 kV in submicron BiCMOS processes [Amerasekera92].

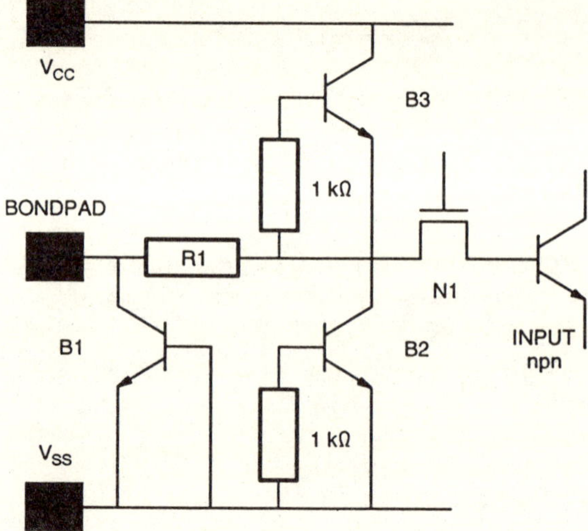

Figure 4.51 Bipolar/BiCMOS input protection scheme.

4.7.5 Layout

Bipolar transistors used in protection circuits must be laid out in straight lines using minimum emitter width. If a suitable base-emitter resistance is used, then the multiple emitter fingers may be used each of minimum width and separated by the minimum spacing allowed. There is no effect of the contact spacing or the collector to emitter spacing in submicron bipolar processes, and no restrictions need to be placed on these parameters. The nMOS transistors and resistors must be laid out in accordance with the requirements discussed in Sections 4.2 and 4.3.

4.7.6 ESD and performance trade-offs

One of the biggest problems confronting the ESD protection circuit designer in bipolar and BiCMOS circuitry is the requirement for high speed. The ideal protection circuit from the operating perspective will have zero capacitance and zero resistance. This can be achieved to some degree in self-protecting bipolar output buffers, but the need for a minimum length and an emitter-base resistance will affect the speed to some extent.

The areas of ESD protection circuits described here are close to the minimum possible in terms of the number of elements used and the efficiency of the circuit. The easiest solution is to trade-off ESD performance with speed. A 4 kV ESD performance is not always necessary, and the circuit can be fine-tuned to obtain the minimum acceptable ESD level by reducing the size of the protection circuit. Since most of the capacitance comes from the vertical region of the collector–substrate junction, reducing the length of the collector and increasing the emitter area will reduce capacitance and improve ESD levels. The emitter area should be increased by adding emitter fingers of the minimum width in order to avoid emitter crowding effects associated with large area emitters at high current levels. However, this technique will only work if the extrinsic emitter-base resistance is high enough that turn-on of all the emitters is achieved, i.e. the current flow is entirely vertical.

4.8 SUMMARY

In this chapter several protection devices were individually described in detail. Each of these has a limited applicability that is dependent mainly on the process technology. For example, the thick field oxide device worked very well for technologies with feature sizes greater than 1 μm but has not performed as well in advanced processes.

A design synthesis has been shown whereby a total protection scheme can be designed by combining primary protection devices with the secondary protection devices. The SCR device can form a very effective primary protection device for inputs. When its trigger voltage is in the 40–50 V range the secondary protection design becomes critical. But, as shown in this chapter, if the trigger level can be reduced the protection efficiency improves. Both n^+-diffusion and n-well resistors can be used as the isolation elements. Some of the advantages and disadvantages of these resistors have been discussed here.

A protection circuit with low RC delay can also be designed with a polysilicon resistor. Although in the past polysilicon resistors have not been found to be very effective, the modified low trigger SCR described here can make this into an efficient protection scheme. Once a successful protection design is obtained with the polysilicon resistor, it has several advantages over the diffusion resistor including the improvement in RC delay. Some of these features make it attractive for analog circuit or high speed circuit applications.

Design issues related to bipolar and BiCMOS protection circuits have also been discussed. These circuits use the vertical *npn* transistor as the primary protection device. In BiCMOS circuits, the trigger voltage of the *npn* transistor needs to be reduced to below that of the nMOS transistor in order for the protection circuit to be effective. Some schemes are discussed for reducing the trigger voltage using active trigger elements. The issues related to the trade-off between ESD performance and circuit speed, which is important for high speed circuits, are also briefly addressed.

REFERENCES

[Abderhalden91]   J. Abderhalden, *Untersuchungen zur Optimierung von Schutzstrukturen gegen elektrostatische Entladungen in integrierten CMOS-Schaltungen*, PhD Thesis, Eidgenossische Technische Hochschule (ETH), Zurich, 1991.

[Amerasekera90]   A. Amerasekera, L. Roozendaal, J. Abderhalden, J. Bruines, L. Sevat, 'An Analysis of Low Voltage ESD Damage in Advanced CMOS Processes', in *Proc. 12th EOS/ESD Symposium*, p. 143–149, 1990.

[Amerasekera92]   A. Amerasekera, A. Chatterjee, 'An Investigation of BiCMOS ESD Protection Circuit Elements and Applications in Submicron Technologies', in *Proc. 14th EOS/ESD Symposium*, p. 265–276, 1992.

[Avery83]   L. Avery, 'Using SCR's as Transient Protection Structures in Integrated Circuits', in *Proc. 5th EOS/ESD Symposium*, p. 177–180, 1983.

[Carbajal92]   B. Carbajal, R. Cline, B. Andresen, 'A Successful HBM ESD Protection Circuit for Micron and Sub-Micron Level CMOS', in *Proc. 14th EOS/ESD Symposium*, p. 234–242, 1992.

[Chatterjee91A]   A. Chatterjee, T. Polgreen, 'A Low-Voltage Triggering SCR for On-Chip Protection at Output and Input Pads', *Elec. Dev. Lett.*, EDL-12, p. 21–22, 1991.

[Chatterjee91B]   A. Chatterjee, T. Polgreen, A. Amerasekera, 'Design and Simulation of 4 kV ESD Protection Circuit for a 0.8 μm BiCMOS Process', in *Tech. Dig. IEDM*, p. 913–916, 1991.

[Chen88]   K.-L. Chen, 'Effect of Interconnect Process and Snapback Voltage on the ESD Failure Threshold of NMOS Transistors', in *Proc. 10th EOS/ESD Symposium*, p. 212–219, 1988.

[Corsi93]   M. Corsi, R. Nimmo, F. Fattori, 'ESD Protection of BiCMOS Integrated Circuits Which Need to Operate in the Harsh Environments of Automotive or Industrial', in *Proc. 15th EOS/ESD Symposium*, p. 209–214, 1993.

[Dabral93]   S. Dabral, R. Aslett, 'Designing On-Chip Power Supply Coupling Diodes for ESD Protection and Noise Immunity', in *Proc. 15th EOS/ESD Symposium*, p. 239–250, 1993.

[Dabral94]   S. Dabral, R. Aslett, T. Maloney, 'Core Clamps for Low Voltage Technologies', in *Proc. 16th EOS/ESD Symposium*, p. 141–148, 1994.

[DeChiaro86]   L. F. De Chiaro, 'Input ESD Protection Network for Fineline NMOS – Effects of Stressing Waveform and Circuit Layout', in *Proc. 24th IRPS*, p. 206–214, 1986.

[Diaz94]   C. Diaz, G. Motley, 'Bi-Modal Triggering for LVSCR ESD Protection Devices', in *Proc. 16th EOS/ESD Symposium*, p. 106–112, 1994.

[Duvvury83]   C. Duvvury, R. N. Rountree, L. S. White, 'A Summary of Most Effective Electrostatic Protection Circuits for MOS Memories and Their Observed Failure Modes', in *Proc. 5th EOS/ESD Symposium*, p. 181–184, 1983.

[Duvvury85]   C. Duvvury, R. Rountree, D. Baglee, A. Hyslop, L. White, 'ESD Design Considerations for ULSI', in *Proc. 7th EOS/ESD Symposium*, p. 45–48, 1985.

[Duvvury86]   C. Duvvury, R. McPhee, D. Baglee, R. Rountree, 'ESD Protection Reliability in 1-μm CMOS Circuit Performance', in *Proc. 24th IRPS*, p. 199–208, 1986.

[Duvvury87]   C. Duvvury, R. N. Rountree, Y. Fong, R. A. McPhee, 'ESD Phenomena and Protection Issues in CMOS Output Buffers', in *Proc. 25th IRPS*, p. 174–180, 1987.

[Duvvury88]   C. Duvvury, R. Rountree, 'Output ESD Protection Techniques for Advanced CMOS Processes', in *Proc. 10th EOS/ESD Symposium*, p. 206–211, 1988.

[Duvvury89]   C. Duvvury, T. Taylor, J. Lindgren, S. Kumar, 'Input Protection Design for Overall Chip Reliability', in *Proc. 11th EOS/ESD Symposium*, p. 190–198, 1989.

[Duvvury90]   C. Duvvury, 'ESD Reliability For Advanced CMOS Technologies', in *Int. Elec. and Dev. and Mat. Symp.*, p. 265–272, 1990.

[Duvvury92]   C. Duvvury, C. Diaz, 'Dynamic Gate-Coupled NMOS for Efficient Output ESD Protection', in *Proc. 30th IRPS*, p. 141–150, 1992.

[Fukuda88]   Y. Fukuda, K. Kato, 'VLSI ESD Phenomenon and Protection', in *Proc. 10th EOS/ESD Symposium*, p. 228–234, 1988.

[Hulett81]   T. V. Hulett, 'On-Chip Protection of NMOS Devices', in *Proc. 3rd EOS/ESD Symposium*, p. 90–96, 1981.

[Keller81]   J. K. Keller, 'Protection of MOS Integrated Circuits from Destruction by Electrical Discharge', in *Proc. 3rd EOS/ESD Symposium*, p. 73–79, 1981.

[Ker92]   M.-D. Ker, C.-Y. Wu, C.-Y. Lee, 'A Novel CMOS ESD/EOS Protection Circuit With Full-SCR Structures', in *Proc. 14th EOS/ESD Symposium*, p. 258–264, 1992.

[Krakauer94]   D. Krakauer, K. Mistry, 'Circuit Interactions during Electrostatic Discharge', in *Proc. 16th EOS/ESD Symposium*, p. 113–119, 1994.

[LeBlanc91]   J. P. LeBlanc, M. D. Chaine, 'Proximity Effects of Unused Output Buffers on ESD Performance', in *Proc. 29th IRPS*, p. 327–330, 1991.

[Lin93]   D. L. Lin, 'ESD Sensitivity and VLSI Technology Trends: Thermal Breakdown and Dielectric Breakdown', in *Proc. 15th EOS/ESD Symposium*, p. 73–82, 1993.

[McPhee86]   R. McPhee, C. Duvvury, R. Rountree, H. Domingos, 'Thick Oxide ESD Performance under Process Variations', in *Proc. 8th EOS/ESD Symposium*, p. 173–179, 1986.

[Palella85]   A. Palella, H. Domingos, 'A Design Methodology for ESD Protection Networks', in *Proc. 7th EOS/ESD Symposium*, p. 169–174, 1985.

[Polgreen89]   T. Polgreen, A. Chatterjee, 'Improving the ESD Failure Threshold of Silicided nMOS Output Transistors by Ensuring Uniform Current Flow', in *Proc. 11th EOS/ESD Symposium*, p. 167–174, 1989.

[Rieck89]   G. Rieck, R. Manely, 'Novel ESD Protection for Advanced CMOS Output Drivers', in *Proc. 11th EOS/ESD Symposium*, p. 182–189, 1989.

[Rountree85]   R. N. Rountree, C. L. Hutchins, 'NMOS Protection Circuitry', *IEEE Trans. Elec. Dev.*, ED-32, p. 910–917, 1985.

[Rountree88]   R. Rountree, C. Duvvury, T. Maki, H. Stiegler, 'A Process-tolerant Input Protection Circuit for Advanced CMOS Processes', in *Proc. 10th EOS/ESD Symposium*, p. 201–211, 1988.

[Scott86]   D. Scott, G. Giles, J. Hall, 'A Lumped Element Model for Simulation of ESD Failures in Silicided Devices', in *Proc. 8th EOS/ESD Symposium*, p. 41–47, 1986.

[Tandan94]   N. Tandan, G. Conner, 'ESD Trigger Circuit', in *Proc. 16th EOS/ESD Symposium*, p. 120–124, 1994.

[Voldman92]   S. Voldman, V. Gross, M. J. Hargrove, J. M. Never, J. A. Slinkman, M. P. O'Boyle, T. S. Scott, J. J. Delecki, 'Shallow Trench Isolation Double-Diode Electrostatic Discharge Circuit and Interaction with DRAM Output Circuitry', in *Proc. 14th EOS/ESD Symposium*, p. 277–288, 1992.

[Voldman94]   S. Voldman, G. Gerosa, 'Mixed-Voltage-Interface ESD Protection Circuits for Advanced Microprocessors in Shallow Trench Isolation and LOCOS CMOS Technology', in *Tech. Dig. IEDM*, 1994.

[Weston92]   H. Weston, V. Lee, T. Stanik, 'A Newly Observed High Frequency Effect on the ESD Protection Utilized in a Gigahertz NMOS Technology', in *Proc. 14th EOS/ESD Symposium*, p. 95–98, 1992.

[Wilson87]   D. Wilson, H. Domingos, M. M. S. Hassan, 'Electrical Overstress in nMOS Silicided Devices', in *Proc. 9th EOS/ESD Symposium*, p. 265–273, 1987.

5

FAILURE MODES, RELIABILITY ISSUES AND CASE STUDIES

5.1 INTRODUCTION

In order to design ICs with good ESD performance it is important that the main design and process parameters which influence the behavior of the inputs and outputs under ESD conditions are known. To determine these parameters and understand their significance, it is necessary to determine the electrical and physical failure modes caused by an ESD stress and the relevant physical mechanisms [Rountree85][Duvvury86]. Failure analysis (FA) is essential in the ESD design path. The first part of this chapter will look at some typical failure mechanisms and their electrical signatures in advanced CMOS processes.

In Chapter 4, the details of the various primary and secondary protection devices and how they can be combined to form effective protection circuits were presented and discussed. But the total circuit reliability is a function of numerous parameters when the protection devices are included. These will be described in the next part of this chapter with illustrative examples. An effective ESD protection circuit scheme for a full IC requires that the protection circuits also prevent damage to circuitry beyond the input and output buffers. As an example, for a particular stress combination of the input pin and the V_{dd} supply, the current can flow internally in the chip between the V_{dd} and V_{ss} connections leading to possible internal damage. This type of weakness must be detected by analysis of the failure mechanism involved before a solution is identified. These details are presented in the second part of the chapter with illustrative case studies. For the sake of clarity, some of the protection concepts are repeated here while discussing the reliability phenomena.

5.2 FAILURE MODE ANALYSIS

5.2.1 Failure Analysis Techniques

The identification of the failure locations and the type of failure is essential to the process of designing and debugging ESD protection circuits and solving ESD problems in existing circuits. In this section we will discuss briefly the failure analysis tools and techniques for location of the damage site and identifying the failure mode.

The first and easiest failure analysis tool for identification of failure location is *Liquid Crystal* analysis. The chip (or wafer) is placed on a temperature-controlled chuck and the temperature is raised to about 50°C depending on the type of liquid crystal used. The liquid crystal is applied to the chip, and the device is powered up. Figure 5.1 shows an example of a liquid analysis of an output buffer damaged by ESD. The dark region at the bottom right-hand side of the picture indicates the hot spot where the damage is located.

As Figure 5.1 shows, hot spots in the device denoting possible failure locations appear as dark regions detectable visually through a microscope. The dark regions extend over a large area and this method does not reveal the exact location of the failure. However, it allows the FA engineer to identify the area that he or she needs to focus on in the more detailed analysis that follows. Typically, the liquid crystal method can detect failures of the order of 100 μA or greater, although an experienced engineer can increase the sensitivity to the order of 1 μA through manipulation of the temperature and the liquid crystal itself. Two major advantages of this technique are its low cost and the fact that it is non-destructive.

A FA tool which is gaining increasing popularity is *Light Emission Microscopy*, also known as *Photon Emission Microscopy* or EMMI. This technique uses photon

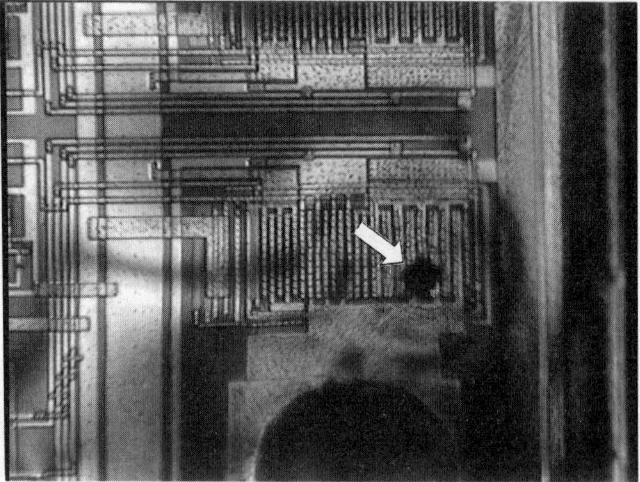

Figure 5.1 Liquid crystal analysis of ESD damage in a CMOS output buffer. The damage is indicated (arrow) by the dark area at the right-hand side of the output transistors.

detectors to identify regions of excessive light emission which signify high current and reverse-biased breakdown [Wills88][Hannemann90]. The device is placed under a microscope and powered up. The image intensifier in the equipment detects the emitted photons and allows visual observation of the emitted light. Figure 5.2 shows a typical failure location in an input protection circuit identified by photon emission. The arrow points to the light emission from the damaged region. The method allows the failure location to be determined with almost pin-point accuracy thereby reducing the analysis time.

The peak of the detector sensitivity is at around 3 eV which makes it ideal for observing photons emitted by reverse-biased junctions. Recent advances in the technology have also made it possible to observe infra-red emission due to thermal effects. The technique does not require any temperature control, and like the liquid crystal method it is non-destructive. However, the equipment is expensive relative to the liquid crystal. It is also possible to use this technique to observe the real-time high current behavior of an ESD circuit which is excellent for understanding the basic phenomena involved [Hannemann90][Amerasekera90][Cavone94]. It must be emphasized that this tool is invaluable in the study of ESD phenomena in ICs and in the design debugging process for ICs [Duvvury92].

Once the failure location has been determined, the device is deprocessed. Deprocessing consists of a series of etches which remove the levels of passivation, interlevel oxide and metal until the silicon is exposed. It is possible to deprocess down to the polysilicon gate before submitting the device to *Scanning Electron Microscopy* (SEM) to examine the damage area. The etch processes tend to remove thermally damaged regions in the silicon leaving the notches or cavities shown later in this chapter in Figures 5.6 and 5.7. Melt regions which extend between the drain and source of the transistor can cause breaks in the oxide and even in the polysilicon (Figure 5.9) and should not be mistaken for oxide breakdown itself. Identifying oxide breakdown is more difficult, and requires very sensitive etches with care being

Figure 5.2 Photon emission analysis of ESD damage in a CMOS input protection. The damage is indicated by the bright spot (arrow) of high photon emission.

taken not to etch through the oxide and remove evidence of damage [Colvin93]. In fact one of the important issues in deprocessing for ESD failures is that the damage can be very close to the surface of the silicon. Hence, the etch processes must be performed in very small increments to ensure that the damaged region is not etched away.

5.2.2 Electrical characteristics after damage

The ESD failure threshold of an IC is determined by monitoring the leakage current, I_{leak}, at the stressed pin. Depending on the failure criterion selected, failure is defined by a change in the leakage current, ΔI_{leak}. It has been shown that ΔI_{leak} in circuits fabricated in an advanced CMOS process can be in one of five categories [Amerasekera92]:

(1) No increase in leakage current, $\Delta I_{leak} = 0$

(2) $\Delta I_{leak} < 10\ \mu A$

(3) $1\ \mu A < \Delta I_{leak} < 100\ \mu A$ (note the overlap with 2)

(4) $100\ \mu A < \Delta I_{leak} < 1\ mA$

(5) $\Delta I_{leak} > 1\ mA$

In conjunction with detailed failure analysis, it has been shown that the above categories can be related to typical failure modes (see Section 5.2.3).

Correlations have been shown between the ESD failure threshold and the median post-stress ΔI_{leak} [Amerasekera90]. Figure 5.3 shows the percentage of failed pins as a function of the ESD stress voltage. These pins were each connected to the output buffer of an IC and each data point is based on the results from 80 output pins. It is seen that although 90% of failures occur above 2000 V, a small percentage fail at ≤ 1000 V. The distribution of the median of the post-stress leakage current with

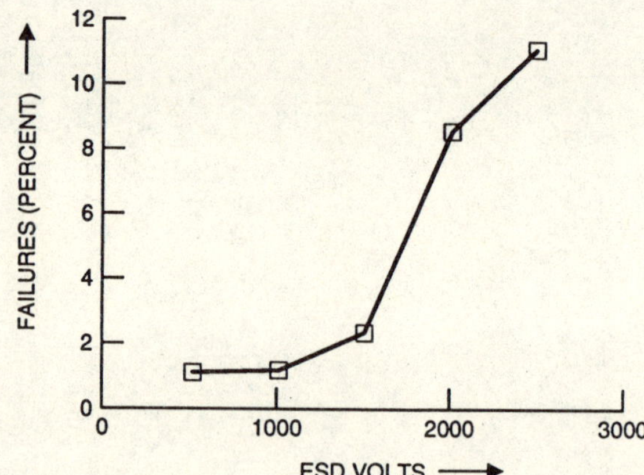

Figure 5.3 Percentage of failed output pins as a function of the applied ESD voltage in a full CMOS IC.

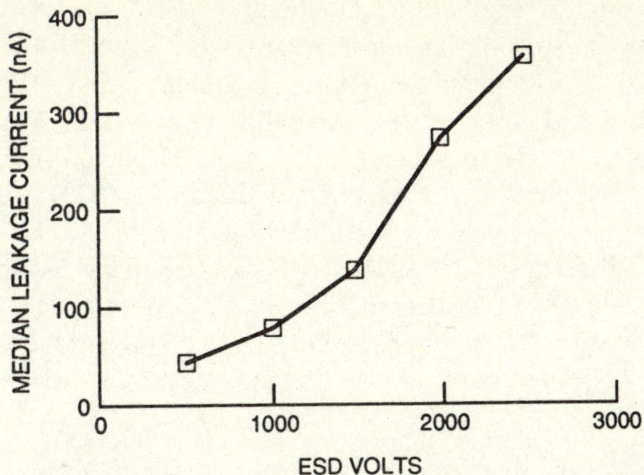

Figure 5.4   Post-stress median leakage current as a function of applied ESD stress.

applied ESD voltage is shown in Figure 5.4. A direct correlation is observed with the ESD failure distribution. A lower median ΔI_{leak} correlates with lower ESD stress voltages. From the distribution of the post-stress electrical characteristic for a given process and design, the main failure mode can be determined.

Device failures due to deviations in the manufacturing process, i.e. *freak* failures, can be identified by monitoring the post-stress leakage currents. The cause of some of these failures may be self-correcting and not require any further action on the part

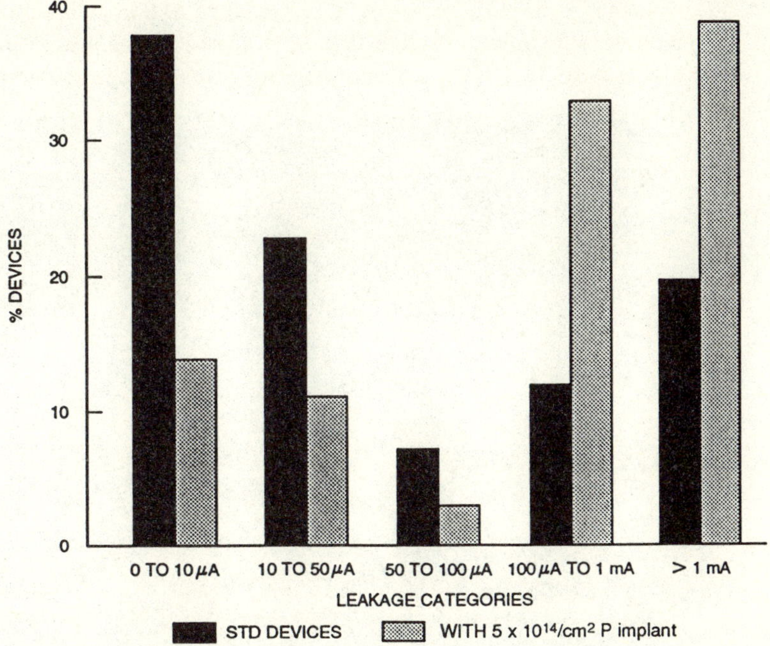

Figure 5.5   Histogram of the distribution of post-stress leakage currents for nMOS transistors after ESD stress at 2 kV.

of the product engineer, while others may indicate that the protection design is too process sensitive. By characterization of the electrical signature associated with ESD damage one is, therefore, able to maintain a control on the process-related ESD behavior and improve the capability to solve ESD problems that arise during development and production. An example of this is shown in Figure 5.5. The histogram shows that with standard processing (STD) the main failure mode is that of *category 1*, while the addition of a $5\times10^{14}/cm^2$ phosphorus implant shifts the main failure mode to *category 5* [Amerasekera92]. The respective ESD failure thresholds moved from a minimum of 500 V for the STD devices to >2000 V for the devices with the additional phosphorus implant as expected from the results of Figures 5.3 and 5.4.

5.2.3 Physical analysis of failure modes

Failure mode analysis has shown that ESD-type failures fall into one of five categories [Amerasekera92].

(1) A common failure mode observed in field oxide devices used as the primary element in input protection circuits is that of holes or notching in the silicon at the field oxide interface as shown in Figure 5.6. The damage is observed at the diffusion connected to the pad and is at a depth related to the bottom of the n^+-diffusion region in the silicon. For this failure mode 100 pA $< \Delta I_{leak} < 10\ \mu$A depending on the stress voltage (500 V to 2 kV).

(2) Damage at the diffusion edge is observed in nMOS transistors used in output buffers or input/output protection as indicated by the arrows in Figure 5.7 and Figure 5.8. $\Delta I_{leak} <10\ \mu$A and the ESD stress voltages range from <500 V to

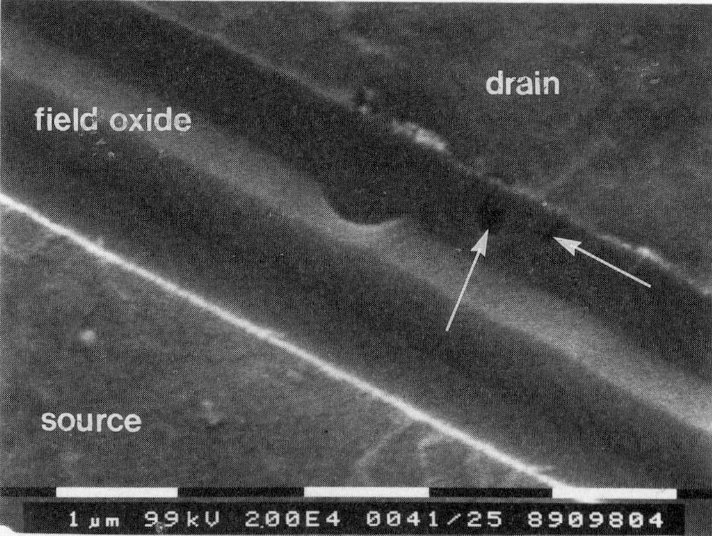

Figure 5.6 SEM photograph showing holes at the silicon to field oxide interface in the drain diffusion of an FOD protection device.

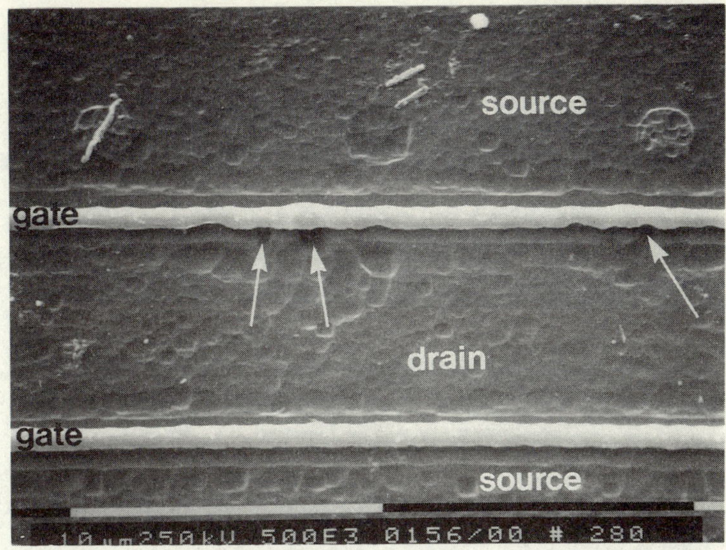

Figure 5.7   SEM photograph of an nMOS transistor in an output buffer showing damage at the drain/gate diffusion edge.

2000 V for this failure mode. This failure mode is most often observed in LDD processes.

(3) Large melt regions due to current filamentation between the n^+ diffusion regions is observed in nMOS transistors (Figure 5.9) and in field oxide devices (Figure 5.10). This is the classic form of ESD damage found in both non-LDD processes and LDD processes. It is the most common failure mode in silicided processes. ΔI_{leak} can be between 1 μA and 100 μA and in extreme cases

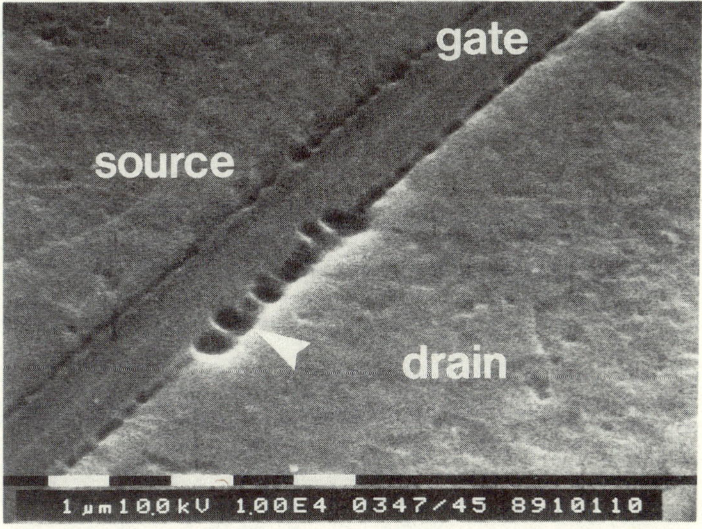

Figure 5.8   SEM photograph of drain/gate diffusion edge damage.

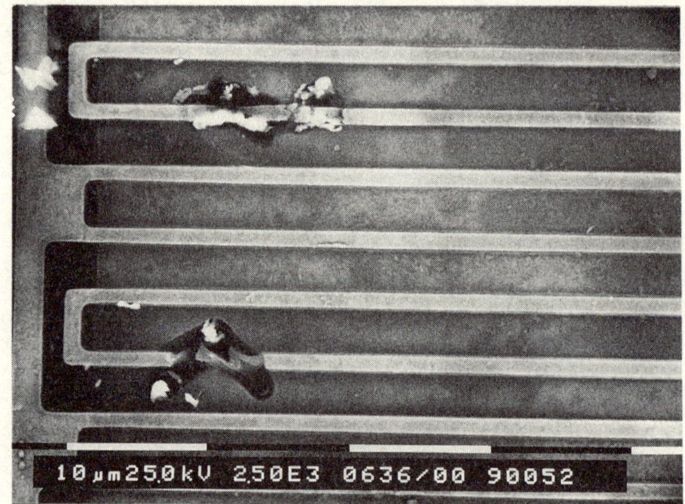

Figure 5.9 SEM photograph of silicon melting due to current filamentation in an nMOS output transistor.

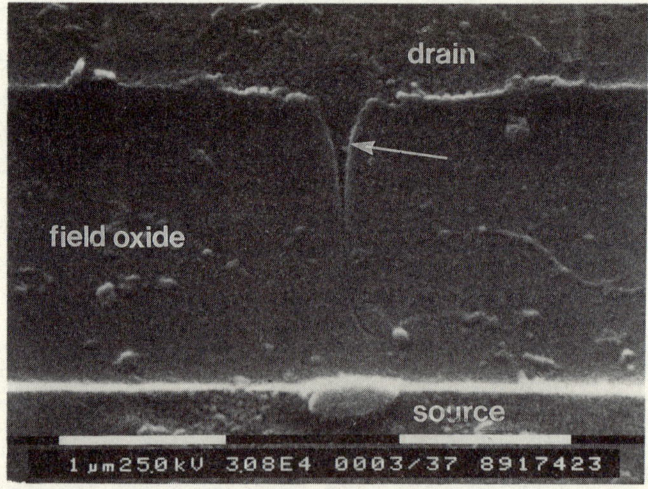

Figure 5.10 SEM photograph of damage due to a polycrystalline filament in a thick oxide device.

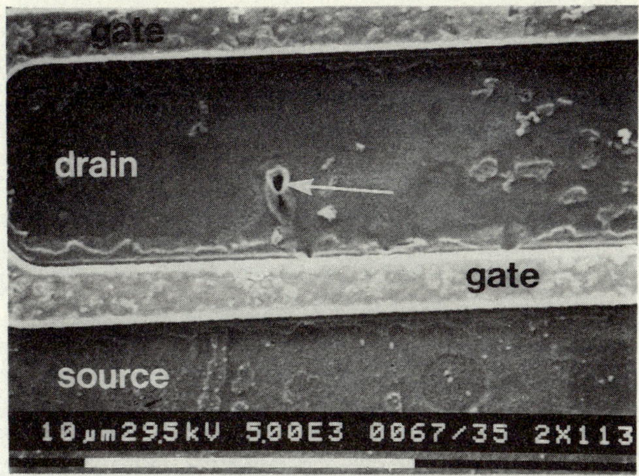

Figure 5.11 SEM photograph of a contact hole in the drain diffusion of an output transistor showing contact spiking.

(especially with silicides) ΔI_{leak} can be >1 mA. The ESD stress levels are very high for this type of damage, usually in excess of 2000 V. However, if the protection or output buffer design is weak then this failure mode can be observed at much lower ESD levels.

(4) Contact spiking in the drain contact region as shown in Figure 5.11 has been observed in both non-silicided [Rountree85] and silicided [Amerasekera92] processes. ΔI_{leak} is usually >1 mA. ESD stress levels are usually in excess of 2000 V before this failure mode is observed, provided the contact to gate spacing (see Chapter 4) has been properly designed. If contact spiking is observed at lower ESD levels it indicates that the contact to gate spacing is not optimized (it could be too small or too large!).

(5) Gate oxide damage in Figure 5.12 with $\Delta I_{leak} \approx 1$ mA is usually observed when the protection circuit has not been properly designed, especially in technologies with gate oxides <175 Å thick.

In addition, examples of polysilicon filaments at the drain edge of nMOS transistors have been shown in Chapter 4. The polysilicon filaments are formed by the migration of polysilicon between the hot junction edge and the gate, aided by the high electric field between the drain and the gate and the temperature gradient (e.g. [Kiefer93]).

Failure modes (1) and (2) are *soft* failure types, where the leakage current changes with additional stressing. The effect is seen in a decrease in the avalanche breakdown voltage, typically about 1 V or 2 V with every additional ESD stress [Amerasekera90][Kuper93]. The post-stress leakage currents have also been observed to decrease after a period of time (24 hours), as well as when subjected to thermal anneal, but they can never be completely eliminated. This indicates that permanent damage has occurred in the junction rather than the occurrence of charge trapping or an increase in surface states. The size and location of the notches in

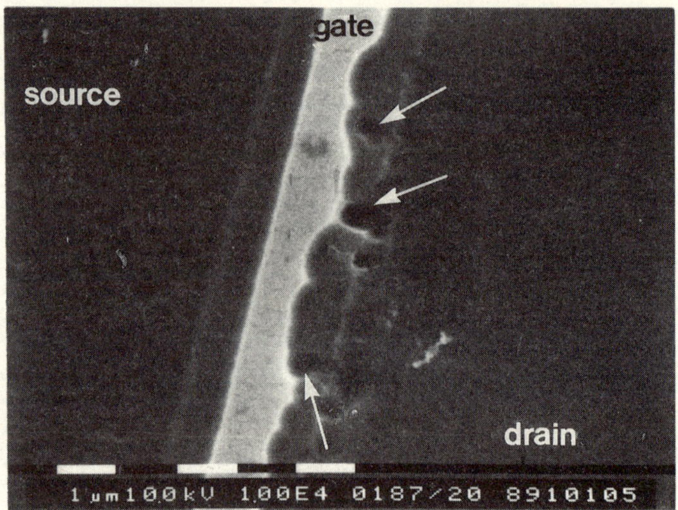

Figure 5.12 SEM photograph of an nMOS transistor showing gate oxide damage.

Figures 5.6 and 5.7 indicates that the damage is in the form of polycrystalline filaments formed thermally at the diffusion edge.

Damage locations are usually in the FOD or FPD nMOS devices. pMOS transistors connected in parallel with the nMOS transistor do not normally show damage because the triggering of the parasitic *npn* is more efficient than that of the parasitic *pnp*. The *npn*, therefore, triggers and shunts all the ESD current and eventually fails, thereby actually protecting the pMOS devices in the circuit. There are conditions under which damage may be observed in the pMOS device as will be discussed later in this chapter.

5.3 RELIABILITY AND PERFORMANCE CONSIDERATIONS

One of the main considerations when designing an ESD protection circuit is to ensure that its incorporation does not alter the intended device performance specifications. For certain pins, the specific function of the pin must also be taken into consideration. In these cases, the input design must be modified so that it will not interfere with the pin applications.

During the qualification of any VLSI chip for production, the high temperature/ voltage burn-in test must be successfully met. If the post-stress characterization results in either *input low* (IIL) or *input high* (IIH) leakage failures, the input protection circuits are usually the cause. The problem is aggravated even more in devices fabricated in advanced process technologies. Such leakage problems are usually associated with non-optimized input protection circuit designs. A detailed analysis would normally reveal the design parameters that need to be optimized. Under some extraordinary circumstances, the problem may not even be associated with the input design but is caused by an interaction of process and protection circuit layout. In such cases, this type of susceptibility to process variations can be corrected by changing the layout while retaining the original protection design.

Here, the various input protection design considerations for good ESD protection, as well as the issues relating to post-burn-in performance reliability, will be considered. The input protection schemes for advanced CMOS technologies and their operation are discussed. The optimum protection circuit design issues for consistent ESD performance are also examined. The post-burn-in reliability issues include both bakeable and non-bakeable leakage failures. The latter type is associated with the ESD protection circuit and will be discussed here. The input design for pins with special applications will be covered with an example. The post-burn-in bakeable leakage phenomena and the solutions to reduce the leakage are also discussed.

Since several different aspects are involved in achieving input protection designs for overall chip reliability, the basic approach in this chapter is to cite the different design experiences and discuss how problems and weaknesses were corrected.

It should be noted here that the protection circuit failures described in this chapter are case studies that actually occurred in product chips. Solving ESD problems involves iterative cycles. For that reason, we have walked through the development of a protection circuit to illustrate the wrong approaches and the corrective actions. Ineffective protection circuits are deliberately considered to show how failures can occur. The main objective in this chapter is to show how potential ESD failure can occur at the pin or in the internal circuits and the use of failure analysis to identify the causes and make improvements for higher ESD performance and better functional reliability in the IC.

5.4 ADVANCED CMOS INPUT PROTECTION

As mentioned previously (Chapter 4) the thick-field device has served as the primary protection circuit for a few generations of nMOS technologies [Keller81] [Hulett81][Duvvury83]. CMOS is now the mainstream of most VLSI chips and the n-channel thick-field device (or FOD) is still effective as long as abrupt junction processes are used. However, for improved reliability and performance, graded junction or lightly doped drain (LDD) transistors with or without silicided diffusions have become important. We also considered the advanced process options and described that the thick-field device is no longer effective and a lateral SCR device is a good option [McPhee86][Rountree88]. In fact, this lateral SCR in a CMOS process can be effective even if LDD junctions and silicided diffusions are not present. An overall protection circuit with the lateral SCR is again first described below.

Consider the cross-section of the LSCR repeated in Figure 5.13. The lateral SCR consists of p^+ and n^+ diffused regions in an n-well connected to the input pad and an n^+ diffused region connected to a V_{ss} common terminal. Thus, a pnp is formed with the input p^+ as emitter, the n-well as base and the p-substrate as collector. Similarly, an npn is formed with the n-well as collector, p-substrate as base, and the n^+ connected to the V_{ss} as emitter. During normal operation the n-well and the emitter of the pnp are tied to the same potential and the pnp does not turn on. The details of the

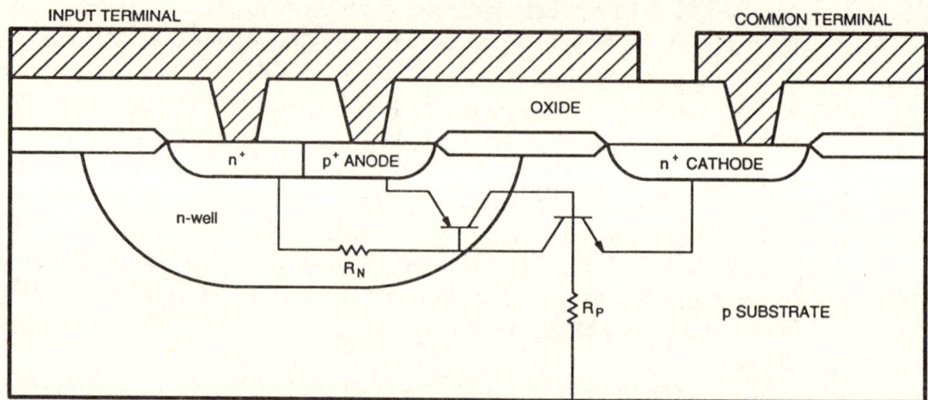

Figure 5.13 Cross-section of the LSCR device.

LSCR trigger and operation during the ESD event have been discussed in Chapter 3. When high current conduction takes place, the total anode-to-cathode potential is only about 1 to 2 V. In general, the holding voltage of the LSCR is determined by the p-substrate resistance R_p and the anode to cathode spacing. The switching or trigger voltage of the LSCR is mainly determined by the n-well overlap of the p^+ anode. A typical value for this parameter varies from 3 μm (for 1.0 μm technologies) to 5 μm (for 2.0 μm technologies). The corresponding trigger voltage can vary from 50 to 70 V. Thus, the SCR devices trigger at higher voltages than the thick-field devices used in the previous protection circuits.

Just as in the case of the thick-field primary protection design, a field plated diode (FPD) and a diffusion resistor are needed to provide the overall protection. The schematic for this is shown in Figure 5.14. The resistor can be either p^+ or n^+ diffusion and both are illustrated in the figure. During an ESD event, the FPD initially clamps the voltage and protects the input buffer gate oxide, while the diffusion resistor limits current to the FPD. With the FPD operating in the breakdown mode as a lateral *npn*, the $I \times R$ drop across the resistor increases the voltage at the pad and eventually leads to the triggering of the SCR device. The triggering voltage or triggering current is determined by the LSCR layout and the process. Since the SCR devices trigger at a higher voltage than the thick-field devices, a more severe stress is placed on the FPD. In fact, the trigger level or the trigger current is basically determined by the secondary protection. This was originally discussed by Rountree *et al* [Rountree88]. A more thorough analysis is presented here.

Consider Figure 5.15 where the cross-section of an input protection with the LSCR is shown. When the FPD goes into the breakdown mode, the hole current in the substrate forward-biases the emitter-base junction of the *npn* in the SCR. This causes the SCR to trigger at a lower voltage than if the SCR triggers itself through avalanche breakdown of the n-well to substrate junction. Thus, the interaction with the FPD is important. In inefficient designs, the SCR will not fire properly and often cause failure at the FPD. An example of this is illustrated in Figure 5.16.

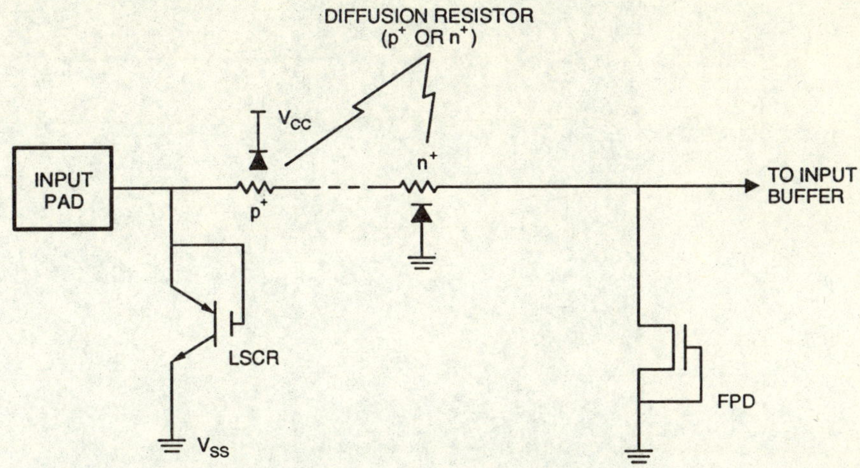

Figure 5.14 Input protection schematic with the LSCR as the primary protection and a field plate diode with a diffusion resistor as the secondary protection. The diffusion can be either n^+ or p^+ as shown.

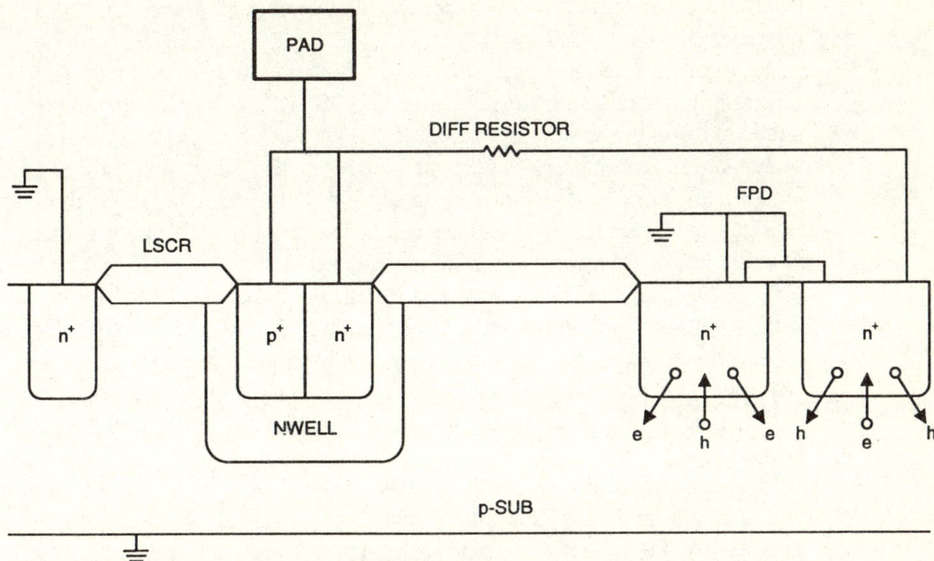

Figure 5.15 Composite input protection circuit using an SCR and an FPD. Holes injected into the substrate due to avalanche breakdown in the drain of the FPD can aid in triggering the SCR device.

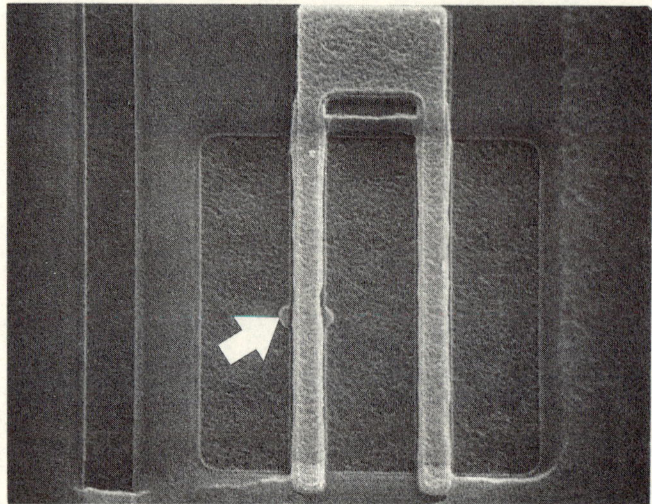

Figure 5.16 Damage to the FPD due to positive polarity ESD stress between pad and V_{ss}. This damage mode of poly gate melt filament shorting to drain occurred since the SCR did not trigger properly.

5.5 OPTIMIZING THE INPUT PROTECTION SCHEME

The importance of the proper design for secondary protection is illustrated through an example in this section. Detailed analyses of this type are necessary to understand the input protection operation and the correct measures needed to improve its efficiency.

An input protection design with the LSCR is used as an example. The process technology for this structure is 1 μm CMOS with LDD drain/source junctions and silicided diffusions. A cross-section of this protection circuit is shown in Figure 5.17. Note that in this scheme the isolation resistor is of p^+ diffusion. In the same n-well as the resistor, an n^+ diffusion is connected to V_{cc} to form a lateral diode. The diode function is to suppress the voltage levels above V_{cc} which might occur at the input. During routine testing, this protection failed below 2 kV of stress. It was suspected that the SCR was not triggering and its effectiveness was tested by disconnecting it from the protection circuit using a laser cutter. The FPD plus resistor combination is then stressed on its own. Step-stressing before and after the laser cut revealed the same average protection level at 1200 V for the Human Body Model (HBM) test with respect to V_{ss}. The results are shown in Table 5.1 and are indicated as *standard process*.

In a subsequent test when the starting stress level was 3 kV, the same protection circuits with the SCR left intact passed 6 kV! Therefore, these same protection circuits that failed 2 kV passed higher stress levels. To understand these results, a more detailed study was done (with the SCR intact) by applying positive stress with respect to V_{ss} from both the low and high ends. For the low-end stressing, the starting voltage was 100 V with incremental steps of 100 V and for the high-end

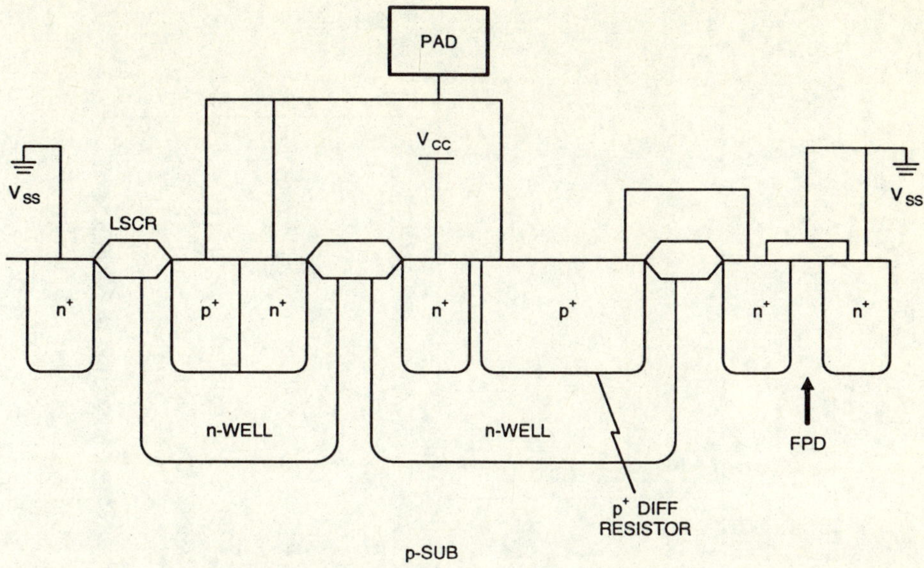

Figure 5.17   Composite input protection scheme with LSCR, p^+ diffusion resistor/ diode, and FPD.

stressing, the starting voltage was 4 kV with decreasing steps of 100 V. The results are shown in Figure 5.18. Note that the failures occurred between 1100 and 1300 V for the low-end stressing and between 1900 and 2300 V for the high-end stressing. Thus, there is a window between 1100 and 2300 V where the SCR does not trigger and provide protection. Below 1100 V, the protection is provided by the secondary device. A similar analysis is done for stress with respect to V_{cc} and the window in this case was 1900 to 2300 V. These results are shown in Figure 5.19. For successful triggering of the SCR, the minimum protection required from the secondary stage is 2300 V and is consistent from both Figures 5.18 and 5.19. Above 2300 V the SCR triggers and shunts most of the current, thus protecting the FPD/resistor combination. Devices of this type would then show a false failure level for a go/ no-go type of stress test depending on the chosen stress level. For example, these

Table 5.1   Input protection performance for processes with 900 Å and 600 Å titanium.

PROCESS	RESISTOR VALUE	SCR TRIGGER CURRENT (mA)	Vf (HBM) +/V$_{SS}$	Vf (HBM) +/V$_{cc}$
STANDARD (900 Å-Ti)	90 Ω	180	1250 ± 100V	1250 ± 100V
EXPERIMENTAL (600 Å-Ti)	180 Ω	150	>6 kV	>6 kV

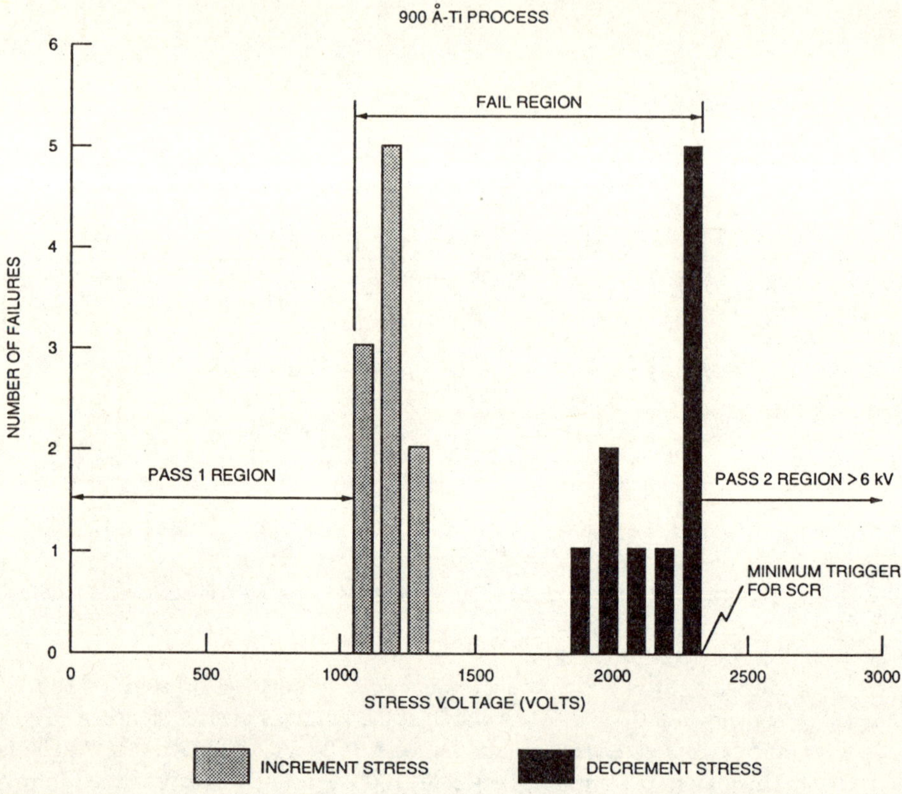

Figure 5.18 Input ESD interim failures for positive ESD stress to V_{ss}. Note that the failures occur in the region where the SCR trigger is critical.

might be guaranteed for 4 kV ESD stress but would be susceptible to common 2 kV stress levels with human handling, such as during a burn-in test setup sequence, and end up as input leakage failures. Of course, if the handling stress levels are much higher (>3 kV) there would be no problem.

The failure analysis and the corresponding leakage failures were characterized for these devices with the results shown in Figure 5.18. For the low-end failures the damages were in the FPD, similar to Figure 5.16. The leakage mechanism is IIH and is indicated as IIH_1 in Figure 5.20. On the other hand, for the high-end stress failures, the damage was in both the p^+ resistor and the FPD and leads to IIH and IIL leakage failures. These are IIH_1, IIH_2, and IIL in Figure 5.20. The damage to the p^+ resistor, which was shorted to the n-well, would lead to IIL leakage because during this test, V_{cc} was high and 0 V was applied at the pad. Referring to Figure 5.20, the shorted p^+ resistor (to n-well) can also short to the nearest p^+ substrate contact causing the IIH_2 failure. The stress current path and the corresponding physical damage are shown in Figure 5.21. Thus, removing this substrate contact from the protection area becomes important. Similar to the V_{ss} stress case, V_{cc} stress failures were also analyzed. As expected, in this case both the low- and high-end failures occurred at

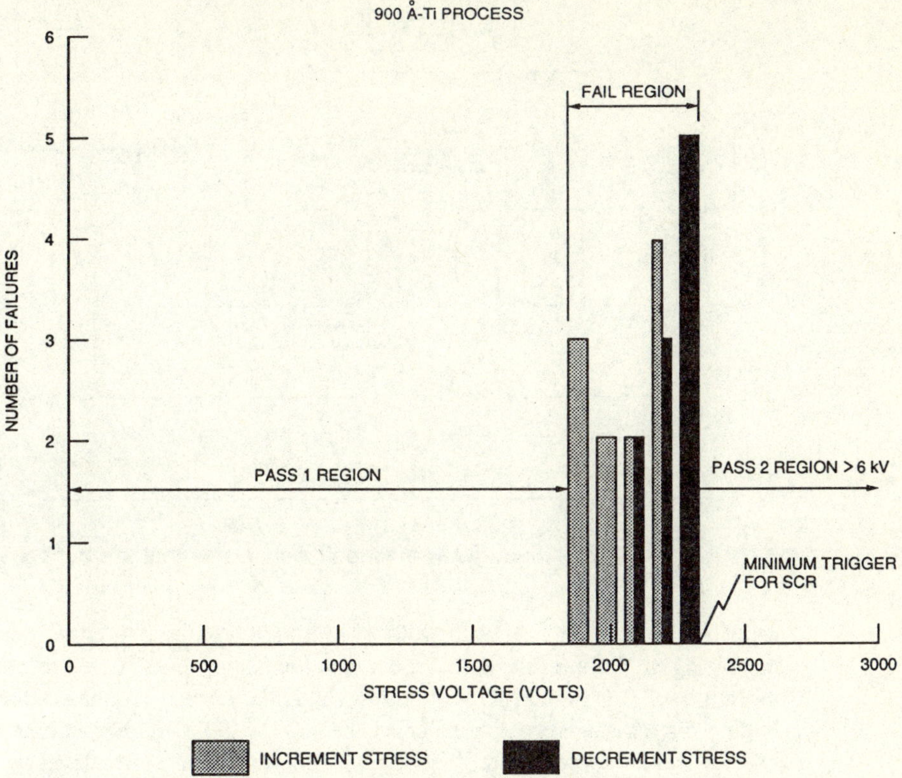

Figure 5.19   Input ESD interim failures for positive ESD stress to V_{cc}. Note that the failures occur in the region where the SCR trigger is critical.

the p^+ resistor and resulted in IIL failures. This makes sense since the FPD was not involved for positive voltage stress to V_{cc}. Thus, with the SCR not triggering properly, the p^+ to n-well diode was easily damaged. The actual stress current path and the observed physical damage for the positive voltage stress to V_{cc} are shown in Figure 5.22. Note that once the p^+ to n-well diode was damaged, the stress current shorts the V_{cc} contact to the V_{ss} contact and can additionally lead to the IIH_2 leakage shown in Figure 5.17.

The above analysis has shown that potential IIL/IIH leakage failures could occur if the input protection is not effective for all stress voltage levels. As an attempt to improve the protection circuit, the same protection circuits were fabricated with a thinner silicide process. The deposited titanium (Ti) for this process was reduced from the standard 900 Å to 600 Å. The results showed that there was no window where the SCR did not fire properly. That is, step stress results showed no failures to above 6 kV (Table 5.1). The same is true for decreasing stress steps. This must mean that the isolation stage performance has improved with the thinner silicide process. (Similar to the V_{ss} stress, V_{cc} stress also shows good performance to >6 kV.) This improvement in the isolation stage performance was verified by detaching the SCR

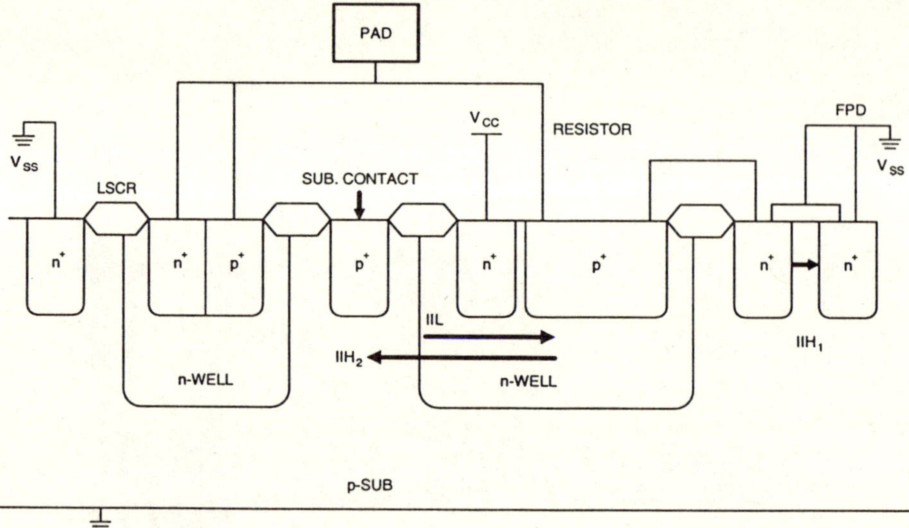

Figure 5.20 Input IIL/IIH leakage mechanisms that are not bakeable.

using a laser cutter for both 900 Å and 600 Å Ti thickness variations and measuring the failure thresholds of the isolation stage alone. These results are shown in Figure 5.23. Note that with the thinner silicide, there is an apparent improvement in the isolation stage protection. This is because with the thinner silicide the isolation resistor value approximately doubled. The minimum ESD level at which the secondary stage was damaged was determined to be 1100 V from the analysis of Figure 5.18. The higher resistance, therefore, requires less current to raise the pad voltage to the trigger level for the SCR, allowing the full protection to be effective at ESD levels <1100 V.

As shown, good input protection levels can be realized in this case by either altering the process or increasing the isolation stage resistor design value to an adequate level. The FPD device width and the resistor value are the main parameters that determine the secondary protection level and consequently the successful

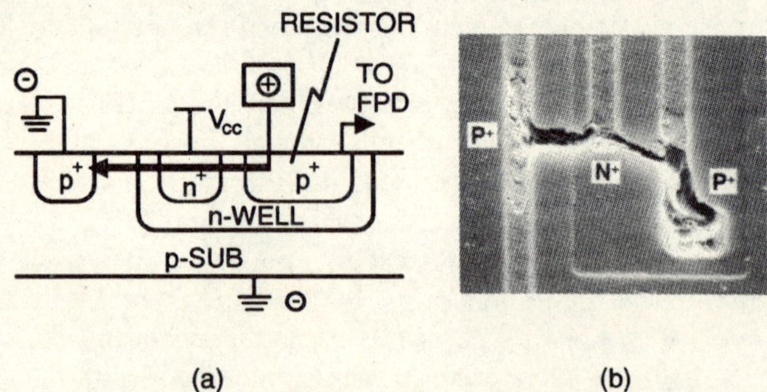

(a)   (b)

Figure 5.21 Isolation stage damage for low-end stress failure with positive ESD stress to V_{ss}. Note the damage is to the lateral *pnp*.

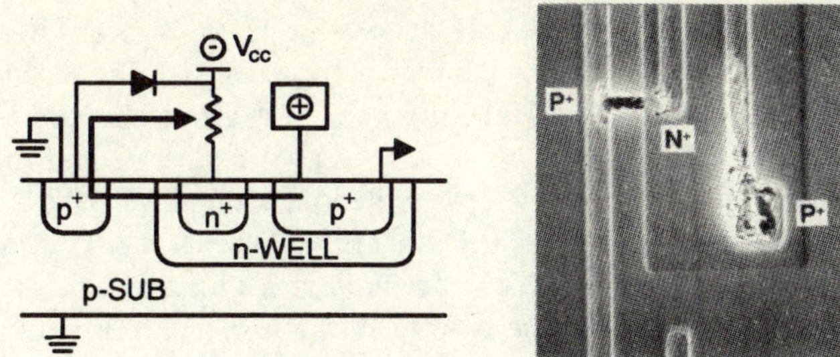

ISOLATION STAGE FAILURES
+/V$_{cc}$ STRESS-DAMAGE THROUGH P$^+$ SUBSTRATE
CONTACT TO V$_{cc}$

Figure 5.22   Isolation stage failure for low-end stress failure with positive ESD stress to V_{cc}. The damage is to the p^+/n-well diode first and then from the V_{cc} contact to the V_{ss} contact.

operation of the SCR. The channel length of the FPD is not found to play any major role and thus a safe minimum length can be chosen without concerns for any subthreshold leakage problems. The FPD channel length, however, can be important for some special application pins. This issue is discussed in Section 5.6.

The protection circuit reliability presented here considered the ineffective triggering with an LSCR. It should be noted that converting the LSCR to an MLSCR, which was described in Chapter 4, could alleviate some of the reliability problems discussed here. This is because the trigger voltage of the MLSCR is lower and is closer to the on-voltage of the lateral *npn* associated with the FPD. Hence, using the MLSCR in the total input protection scheme will provide a more consistent ESD protection circuit.

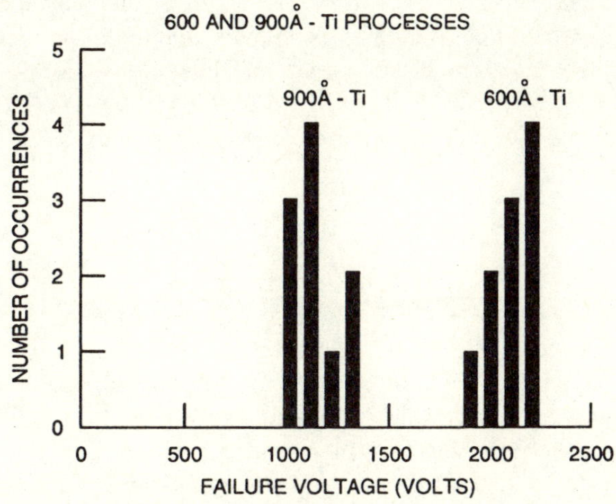

Figure 5.23   Isolation stage performance with the SCR cut off for 900 Å and 600 Å titanium processes. The stress was $+/V_{ss}$.

Some circuit applications such as in automotive or industrial ICs cannot tolerate any accidental triggering of an SCR due to high voltage spikes during normal operation. In these cases the higher triggering LSCR scheme is more desirable but the protection circuit reliability as described here should be carefully evaluated and characterized to ensure optimum ESD protection and reliability.

In another example of input protection design, the layout of the resistor itself can become important for chip reliability. During the implementation of the above protection scheme for silicide technologies a long resistor is required. But because of space constrictions at the bond pad, the resistor must be invariably bent to achieve about 200 Ω. This type of layout could also cause unexpected problems as will be discussed below.

The protection level for a circuit with a bent resistor, as measured using a commercial HBM tester, yielded a high level (>6 kV) of protection. But the chips had a high rate of burn-in loss with IIL and IIH failures. These were found to be non-bakeable and were most likely a result of protection circuit damage. The failure analysis indicated damage to the p^+ resistor as shown in Figure 5.24(A). Any damage to the p^+ resistor can cause leakage of the reverse-biased p^+/n-well diode (Figure 5.20) for IIL failures, or if the damage extends to the n-well/p-substrate junction it can cause an IIH failure. Hence, these burn-in failures are not surprising. This apparent discrepancy between the tester results and burn-in failures is described below.

Analysis of the circuit revealed that at low ESD stress levels ($\approx$1 kV) the current densities in the silicided regions of the bent portions of the p^+ resistor were very high. The silicide was heated beyond its eutectic temperature (i.e. it ceased to exist as a conductive layer) which effectively increased the resistance of the diffusion. The nature of the thermal process is such that no damage was seen outside the resistor (Figure 5.24(B)). Thus, once this self-protection was formed, no degradation of the protection level was seen. But if the starting ESD stress level is 2 or 3 KV, the diode degrades and IIL/IIH failures occur because the damage extends beyond the p^+ resistor surface as shown in Figure 5.24(C). This correlated with the post-burn-in failures where poor handling techniques combined with HBM stress levels of 2 KV or above caused the leakage failures. The solution in this case was first to improve the handling procedures for the ICs during burn-in. A long-term solution was implemented to redesign the protection circuit with a different layout that avoids the bent resistor.

5.6 DESIGNS FOR SPECIAL APPLICATIONS

The design and reliability issues for special applications can be understood by considering actual product chips as they go through different process cycles. For several reliability and functionality factors the process is sometimes changed. In the example given here the process is changed from LDD junctions to abrupt junctions. The reader must keep this in mind while understanding how ESD circuits can cause reliability problems.

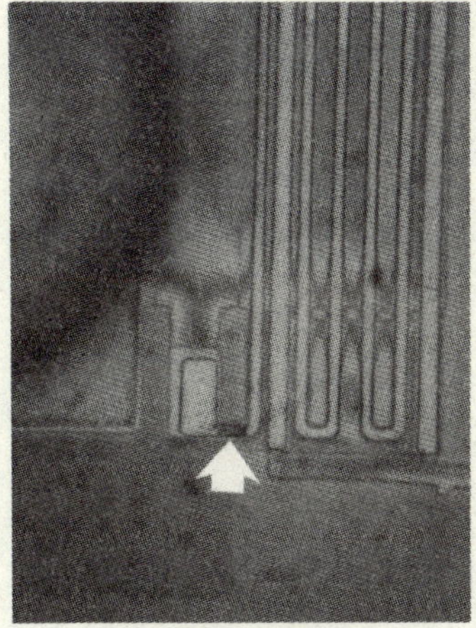

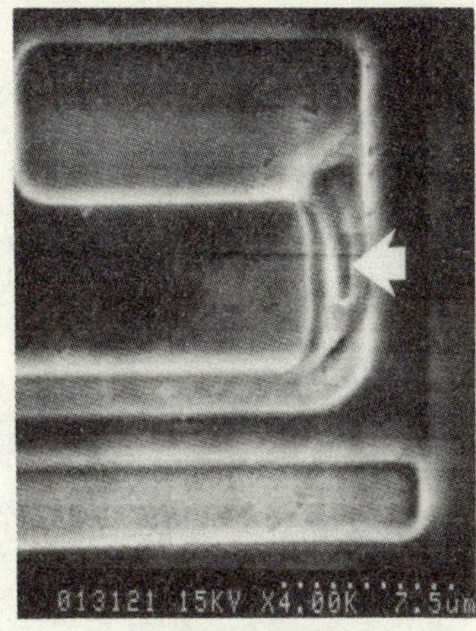

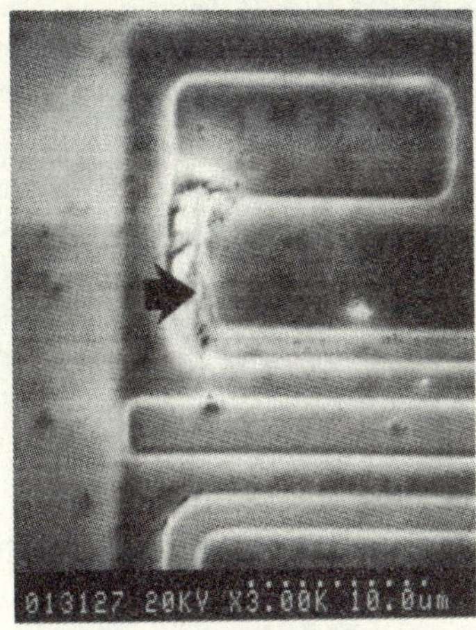

Figure 5.24   (A) Damage to the p^+ resistor at the bend in the layout. (B) p^+ resistor damage internal to the resistor which does not result in IIL or IIH leakages. (C) p^+ resistor damage internal and external to the resistor which results in both IIL and IIH leakages.

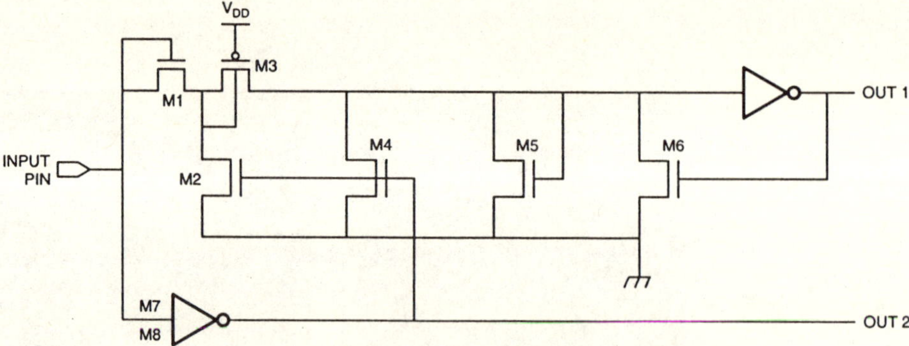

Figure 5.25 Tri-level detection circuit for the HV pin.

Table 5.2 Decode logic states for the tri-level detector circuit.

INPUT PIN (VOLTS)	OUT 1 (BOOLEAN)	OUT 2 (BOOLEAN)
0V	1	1
5V	1	0
>7V	0	0

The input protection circuits are usually optimized for the best possible performance without any leakage concerns. It is usually recommended that the channel length of the FPD should be kept at a minimum since this will improve the performance of the lateral *npn* transistor. Also, it has been reported [Maloney88] that shorter channel lengths are more beneficial for the Charged Device Model stress. However, a minimum channel length might cause problems for certain pin applications, as shown by the example below.

The specific pin under consideration is a high voltage pin (greater than V_{cc}). A high voltage buffer is not only able to differentiate between the normal logic '0' and '1' states, but also able to detect a third 'HV' (high voltage) level. This feature is used in the example microcontroller chip to place the device into special operating or test modes without dedicating a pin for that purpose. This is extremely useful for in-factory testing or device emulation without compromising the need to minimize the number of input/output (I/O) pins. A typical circuit to achieve the tri-level logic function is shown in Figure 5.25. The circuit consists of special devices capable of handling up to 15 V inputs. This is achieved via double-level polysilicon technology. The voltage level at the input pin is decoded by the circuit into two separate digital bi-level logic levels internal to the device (Out 1 and Out 2 in Figure 5.25). Table 5.2 shows the truth table for the circuit assuming a nominal operating V_{cc} of 5 V.

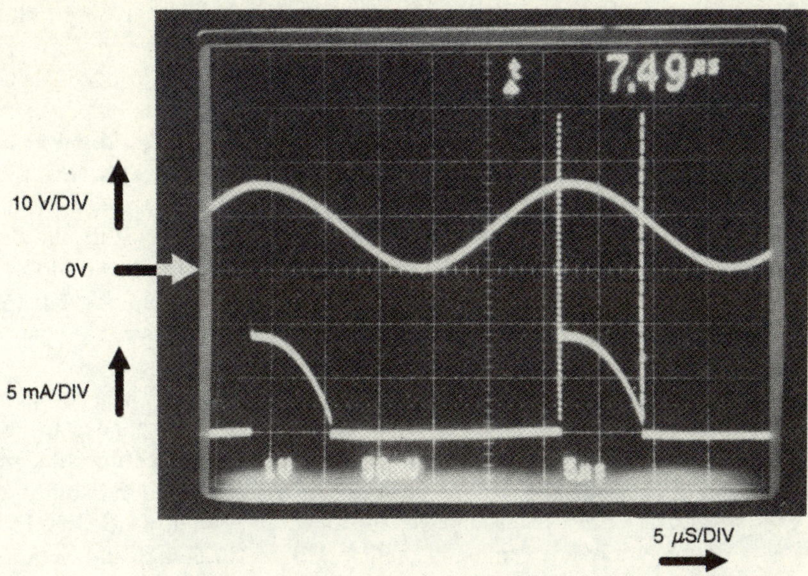

Figure 5.26   The voltage and current waveforms showing the HV pin trigger and snapback phenomena.

The ESD protection circuit used for the HV pin is also the same as for the other inputs consisting of an LSCR, FPD, and a diffusion resistor. This is incorporated in a family of microcontroller chips with different processes. During testing with the HV pin, certain chips are found to exhibit a sudden increase of input current. The input waveform and the current through the pin are shown in Figure 5.26. Note that when the voltage reaches 16 V, there is a sudden increase in the current and this decreases as the voltage falls to below 10 V. Emission microscopy analysis can be used to understand this. In this example we have traced the drop in voltage to the FPD.

The $I{-}V$ characteristics for an FPD device are shown in Figure 5.27. The high current behavior of the FPD has been discussed in detail in Chapter 3. The avalanche breakdown voltage of the drain junction is controlled by the oxide thickness and the

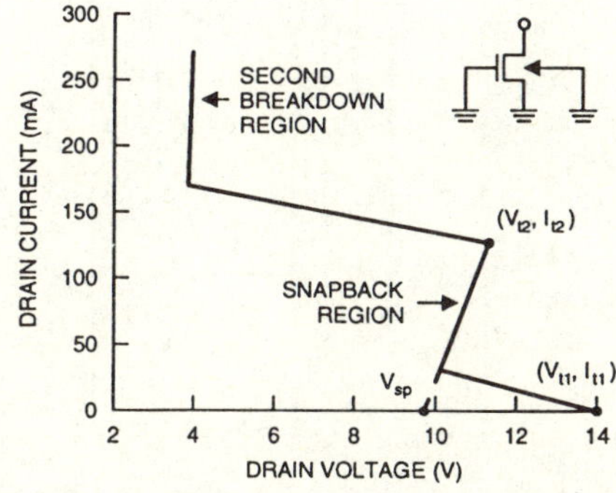

Figure 5.27 $I{-}V$ characteristics for the field plated diode device.

substrate doping. Additionally, the parasitic bipolar holding voltage (or snapback voltage) is controlled by the channel length [Hsu82][Feng86]. The grading of the drain junction also has a significant impact on these voltages. When the voltage at the pin exceeds the trigger voltage the lateral *npn* turns on. Once the bipolar is on it remains in the low impedance mode until the voltage drops below the snapback holding voltage. Thus, the internal node voltage will never be sufficient to allow the internal buffer to recognize it as HV. In addition, the presence of the high current at extended periods of time could cause other damage within the device. To prevent unintentional turn-on of the parasitic bipolar transistor, the FPD must be designed such that the trigger and snapback holding voltages are above the normal operating voltage level. Voltage overshoot and system noise level can also contribute to the problem of unintentional triggering of the protection circuit.

In the case of these microcontroller devices, because of a constant evolution of the ESD protection circuits, different channel lengths happened to be implemented in each design cycle for the FPD device. At the same time, the process fluctuated between LDD and abrupt junction for various other reliability and functionality issues. During this sequence, the operating window for the HV pin drastically changed from one version to another. This is chronologically recorded in Figure 5.28. The required minimum/maximum levels are also indicated as V_{Hmin} and V_{Hmax}. Note that for the 1.6 μm abrupt junction or non-LDD process, the functional failure of the HV pin function (not the ESD performance) is frequent. The maximum is determined by the trigger voltage. The data for this is shown in Figure 5.29. Note again that for the 1.6-μm abrupt junction, the trigger voltage falls below the HV pin maximum. Thus, the FPD channel length needs to be changed so that the trigger voltage is above V_{Hmax}. As a result of this analysis, a simple solution that alleviated this problem was to make the FPD channel length 2.5 μm for the HV pin only. The trigger point of this design is shown by '*' in Figure 5.29. As the trigger voltage

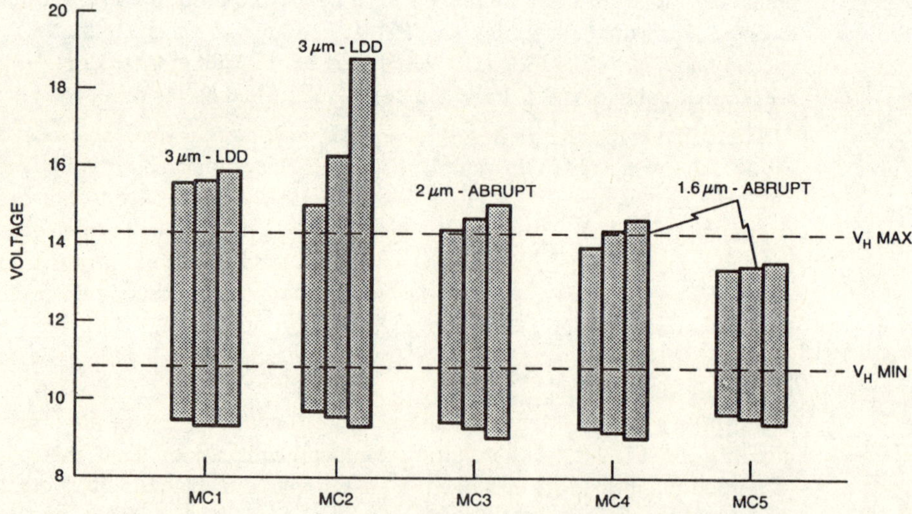

Figure 5.28   HV pin operating window for different processes.

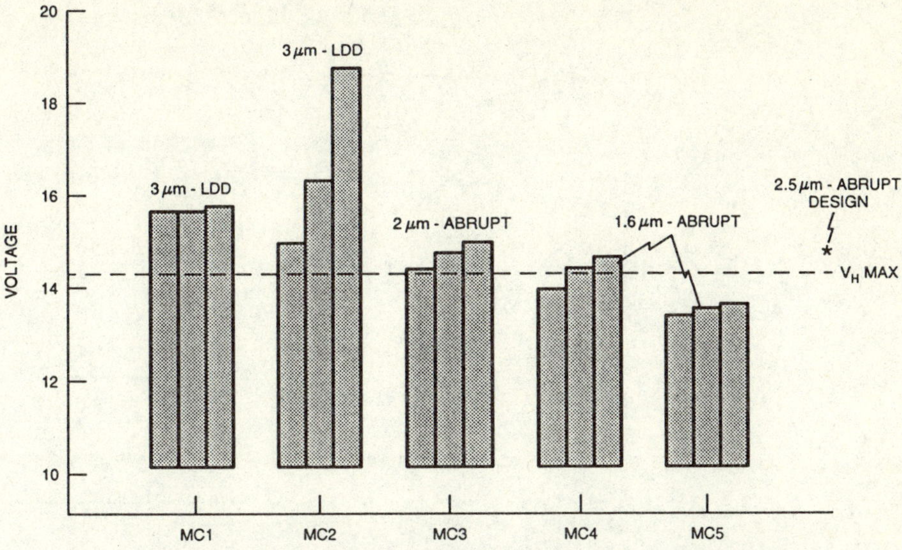

Figure 5.29   FPD trigger voltages for different process options.

increases so does the snapback voltage. But this has little consequence on V_{Hmin} as long as the trigger voltage remains above V_{Hmax}. Even with the increase in snapback voltage the design showed no measurable reduction in ESD performance for this particular pin.

5.7 PROCESS EFFECTS ON INPUT PROTECTION DESIGN

Even effectively designed input protection schemes could cause unusual leakage problems if their layout interacts with process fluctuations. As discussed in the example below, this could lead to IIL/IIH problems again. Process influences on ESD in general are discussed in more detail in Chapter 7.

The leakage problems observed in this case study were post-burn-in IIL/IIH failures that were bakeable and, hence, hard ESD failures could not be suspected. It has been shown that ESD pulses can increase the sensitivity to hot carrier stress degradation [Aur88]. We could assume that ESD-induced hot-carrier degradation in the FPD might be the cause for this bakeable leakage. However, this is not a likely cause since the gate is grounded during normal operation which reduces the chances for hot-carrier injection. Furthermore, leakage at the FPD will not explain why bakeable IIL failures also occurred. In fact, these IIL failures occurred more frequently than IIH failures.

To further investigate this phenomenon, different portions of the input protection circuit were isolated by laser cutting the metal connections. First, the characteristics of a good pin and an IIL leakage pin are compared in Figure 5.30. Note the leakage at 0 V bias at the input. The laser cuts made for the circuit are indicated in

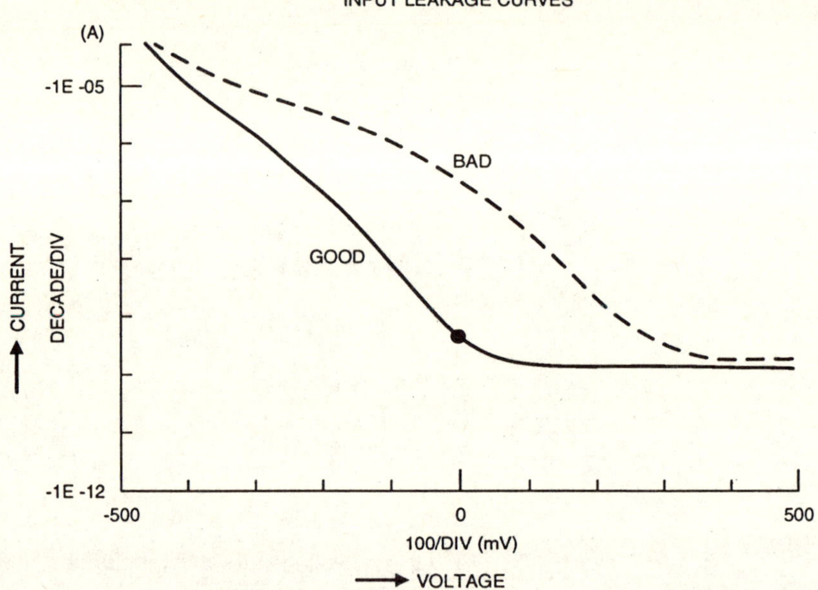

Figure 5.30 Input leakage curves showing good and bad pins. Note the IIL leakage for the bad pin.

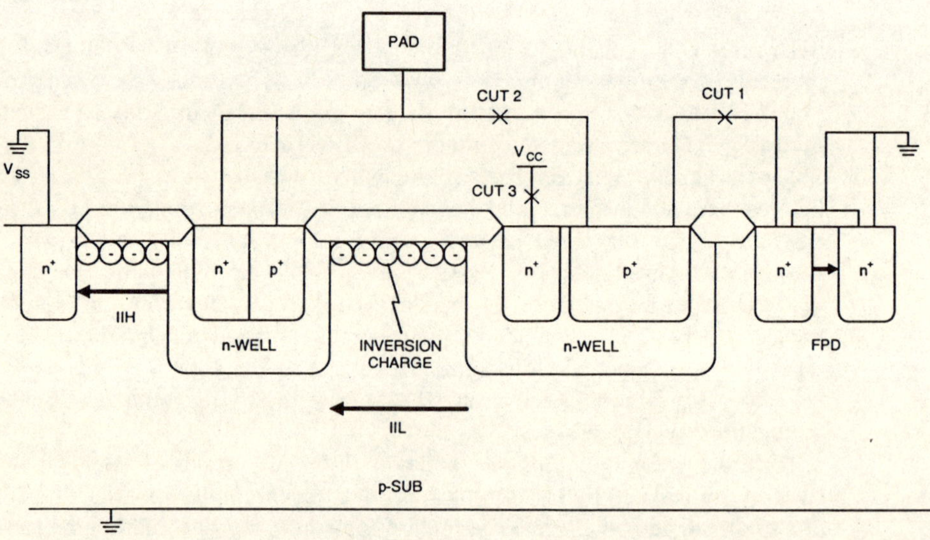

Figure 5.31 Bakeable IIL/IIH leakage mechanisms in an input protection scheme.

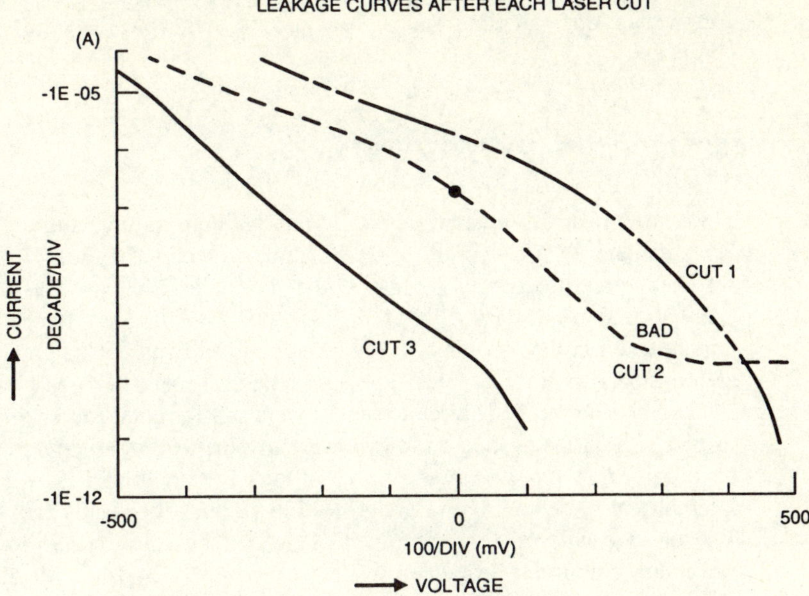

Figure 5.32   Input leakage curves after different laser cuts.

Figure 5.31. The corresponding leakage characteristics are shown in Figure 5.32. After Cut 1, the leakage does not improve and thus there is no contribution from the FPD. The slight increase in the leakage after the cut is caused by laser damage. The leakage essentially remains after Cut 2, meaning that the p^+ resistor is not the cause. Finally, after Cut 3 to the V_{cc} line, the IIL leakage disappears. Thus, the leakage originated from the SCR n-well to the V_{cc} n-well (Figure 5.20) and resulted in IIL failure. The cause for this leakage was traced to small levels of positive mobile ions present in the process at the time. During the burn-in test, since both the wells are under temperature/voltage stress, the mobile ions can easily form an inversion layer underneath the field oxide, causing this bakeable IIL leakage. When 0 V is applied to the input pin during this test, the conduction of the channel from the SCR well to the V_{cc} n-well leads to the increase in leakage. In addition, the bakeable IIH leakages are explained by the fact that, during burn-in, the mobile ions could also form an inversion layer between the n-well and the n^+ cathode of the SCR. The IIH bakeable leakages were less frequently observed. This is because during burn-in charge spreading from both wells contributes to IIL, whereas spreading from one well contributes to the IIH failure.

The obvious solution to the above problem is to identify the source of contamination and decrease the mobile ion level with a cleanup of the process. However, to avoid any potential problems caused by small levels of contaminations that are present even in the cleanest of processes, the protection circuit layout could be modified. That is, the SCR n-well is placed on the other side of the pad, greater than 100 μm away. Alternately, only n^+ resistors can be used but at the expense of not having an effective diode to V_{cc} near the bond pad. With the correct design approaches and an improved process, the bakeable leakage phenomena are eliminated.

5.8 TOTAL IC CHIP PROTECTION

The next part of this chapter considers internal chip failures and protection designs. For complete ESD reliability of an IC chip, protection of the I/O pins alone is not sufficient since there may be many other possible sensitive areas on the chip [Krakauer94]. Internal chip failures could occur even if good protection designs are implemented at the I/O pins. In the remaining sections of this chapter, these issues are considered with examples from microcontroller and DRAM chips.

The input/output ESD circuit requirements call for good protection of the pin with respect to both the ground and the power bus pins. Although effective protection can be designed at the pin, many cases of damage phenomena are known to occur internally in the chip beyond the protection circuit. The issue of protection between V_{dd} and V_{ss} will be first discussed. This will be followed by examples of how protection circuit performance can be sensitive to internal chip layout, independent of its effective design. Several illustrative case studies will be reported to emphasize the internal chip ESD phenomena and their adverse effects.

An effective protection between the power bus lines is often overlooked although this is equally important for overall ESD immunity [Maene92]. A comprehensive method of testing requires not only stressing of every pin with respect to V_{dd} (V_{cc}) and V_{ss}, but also between inputs and outputs. This would lead to very complex internal ESD stress currents, which must be carefully analyzed in order not to compromise the circuit reliability. Consider the overall protection scheme for a CMOS circuit chip that is shown in Figure 5.33. As seen from this diagram, there are several parasitic devices involved in the stress current path for stressing between the different pin combinations of the MIL-STD.

Internal chip ESD damage could result due to direct stress applied between V_{dd} and V_{ss} pins. Some illustrative examples of this and the possible solutions are

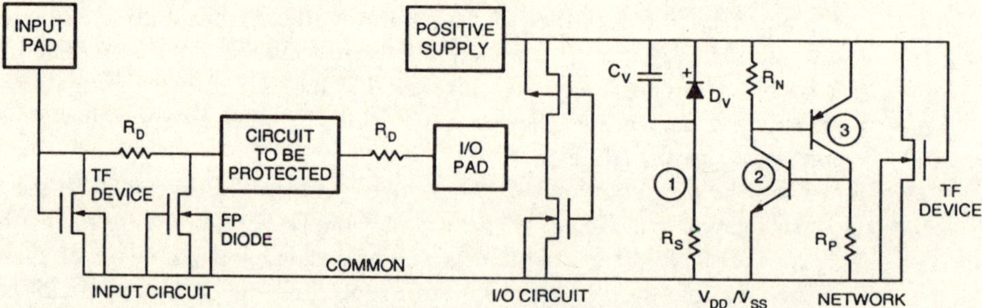

Figure 5.33 Overall CMOS protection scheme, showing basic circuit elements such as Thick Field (TF) devices and Field Plate (FP) diode.

discussed in Section 5.10. Generally, current flow through the internal circuitry can take place when outputs/inputs are stressed with respect to V_{dd} or V_{ss}. Hence, it is important to consider the issues related to the internal circuits and layout, or the overall ESD performance can suffer.

5.9 POWER BUS PROTECTION

Even with effective protection at the pins, many cases of damage phenomena do occur internally in the chip (see e.g. [Duvvury88A][Maene92][Cook93]). Some of these can be directly attributed to inadequate protection provided for stress between the power bus lines while others may be due to deficiencies in the layout of the protection circuits. The power bus protection issues and techniques will be discussed in this section.

The MIL-STD testing method requires stressing between all inputs, outputs, and power bus pins (V_{dd}, V_{cc}) with respect to the ground pin (V_{ss}). Thus it would seem logical to place a protection circuit between V_{dd} and V_{ss} for direct stress between the two [Palella85][Duvvury87B]. This could be a thick field device, shown as 'TF Device' in Figure 5.33. However, it may not necessarily provide the needed protection because there are many parasitic devices in the internal chip that may turn on instead of the designated protection circuit. These parasitic devices essentially form the weak links and reduce the power bus protection. For example, in the internal layout of a chip there are many areas where an n^+ diffusion connected to V_{dd} could be close to another n^+ diffusion connected to V_{ss} such as in the latchup guardrings. These would then form parasitic lateral *npn* devices. Therefore, even if a robust protection circuit is implemented between V_{dd} and V_{ss}, the ESD damage could still occur due to triggering of these parasitic devices which are not designed to sustain high current levels. An example of one such fail site is shown in Figure 5.34. Observe how the molten silicon has tunneled under the polysilicon from the V_{dd} diffusion to the V_{ss} diffusion. In order to ensure that the thick field protection device between V_{dd} and V_{ss} is effective, the diffusion to diffusion spacing for such internal parasitic thick field devices should be made longer, reducing (preferably eliminating) the effectiveness of the parasitic *npn* transistor.

In general, an MOS thick-field device between V_{dd} and V_{ss} is adequate for either positive or negative stress. For positive stress on V_{dd} with respect to V_{ss}, the lateral *npn* in breakdown would form the effective protection; for negative stress on V_{dd} with respect to V_{ss}, the forward-biased diode would turn on provided the substrate is also at V_{ss}. If the substrate connection is not the same as the V_{ss} connection during the ESD stress then the lateral *npn* transistor will be required to trigger and carry the stress current. In large microcomputer chips several power bus lines are often used. For these chips, protection circuits should be used between all bus combinations.

The V_{dd} to V_{ss} capacitance can contribute significantly to the ESD performance of an IC. In large circuits, this capacitance can be of the order of 10 nF. A direct stress between V_{dd} and V_{ss} will first have to charge up this capacitance which will slow the rise time of the current pulse and limit the voltage during stress [Duvvury88B].

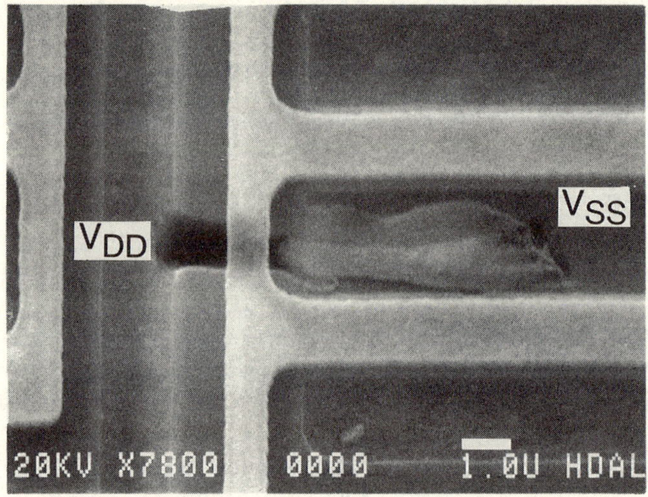

DAMAGE DUE TO V_{DD}-V_{SS} STRESS

Figure 5.34 Damage site observed for V_{dd}–V_{ss} stress.

Triggering of the protection device only occurs after the voltage reaches the required trigger voltage and by this time the stress current could be well below its peak level. Hence, the chip capacitance will limit the stress in the protection device and increase the ESD levels.

As mentioned earlier, a comprehensive ESD testing methodology includes stress between the input pins and V_{dd}. One such case is considered next. Shown in Figure 5.35 is an equivalent circuit schematic of input protection. The thick-field device is shown as a bipolar device for positive stress and as a diode for negative stress. There is no dedicated protection element directly between the pad and V_{dd}. The elements R_V and C_V represent the parasitic resistance and capacitance between V_{dd} and V_{ss}. When the pad is stressed with respect to V_{dd}, the stress current would eventually flow between V_{ss} and V_{dd}. This could lead to damage internally in the chip.

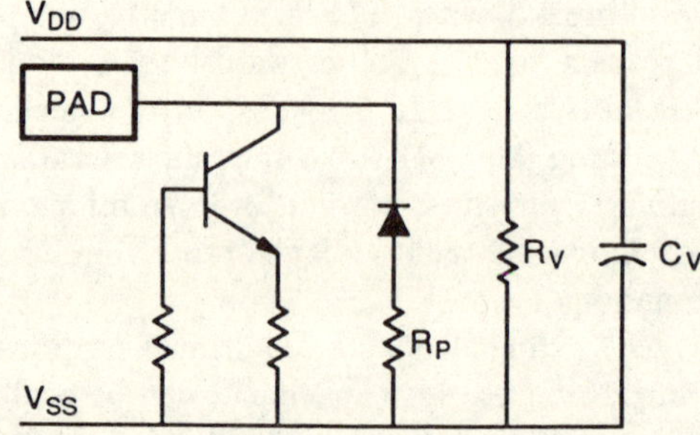

Figure 5.35 Equivalent component schematic of input protection.

5.10 INTERNAL CHIP ESD DAMAGE

In Section 5.9 ESD damage due to direct stress between the V_{dd} and V_{ss} pins was discussed. An example was given where the damage could be directly attributed to V_{dd} and V_{ss} diffusions. However, for applied stress between the power bus pins, some subtle damage phenomena could also result. Furthermore, as mentioned above, for stress between input or output pins and V_{dd}, internal damage could occur, again due to the stress current path between V_{dd} and V_{ss}. Some examples of both cases will be given in the next two subsections.

5.10.1 V_{dd}–V_{ss} Stress current damage

Internal damage due to stress current between V_{dd} and V_{ss} is often difficult to identify if it is not obvious as was shown in Figure 5.34. Liquid crystal analysis can be used to locate these failure sites.

In a CMOS chip the damage site observed through liquid crystal analysis is shown in Figure 5.36(A). In this case a positive ESD stress is applied to the V_{dd} pin with respect to the V_{ss} pin. This damage site is identified as a CMOS inverter in the internal circuitry. Since V_{dd} is stressed positive with respect to V_{ss}, the stress current passes through a parasitic *pnpn* device, as illustrated in Figure 5.36(B), and caused the failure. The parasitic *pnpn* device is formed by the p^+ source diffusion of the pMOS transistor connected to V_{dd}, the n-well, the p^+ substrate, and the n^+ source diffusion connected to V_{ss} of the nMOS transistor. Referring to Figure 5.36(A), it should be noted that the damage extends from p^+ to n^+ through the n^+ guardring contact. One way to reduce susceptibility to this problem would be to increase the resistance (R_G) in the current path. An increase in R_G will reduce the holding voltage

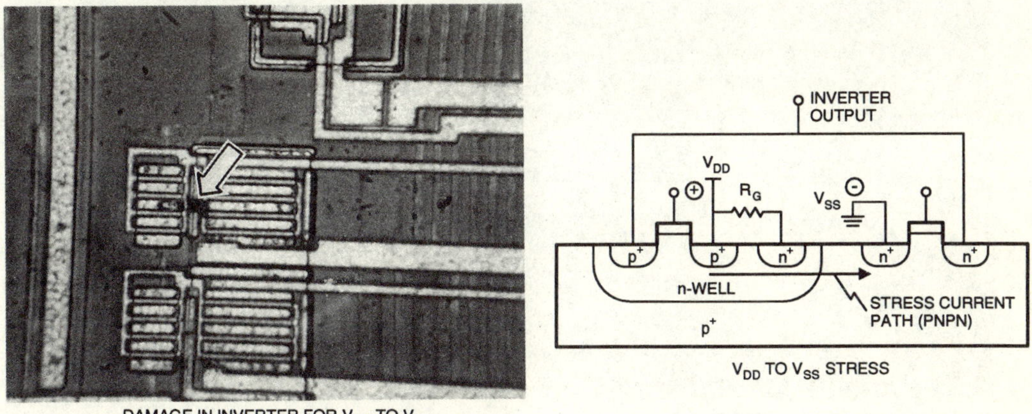

Figure 5.36   Damage site observed in an output buffer for positive stress on V_{dd} with respect to V_{ss}. The *p*-channel device is on the left side of the schematic.

of the parasitic SCR resulting in less power dissipation and heat generation. This could be achieved by identifying the inverter fail sites and removing the guardring contacts in the direct current path. However, the latchup performance of the CMOS chip must not be affected by this approach. For this reason only the guardring contacts facing the nMOS device are removed. Test structure analysis has shown a marked improvement in the ESD performance of individual buffer devices without the direct path guardring contacts (4 kV for Human Body Model stress), as compared to those with complete guardring contacts (600 V for Human Body Model stress). In both structures the same failure mechanism occurred but the failure voltage was higher in the first structure.

The example cited here is for the case of an advanced CMOS process involving silicided diffusions. Since it has been reported earlier that silicided diffusions are more susceptible to heat damage due to ESD [McPhee86][Duvvury86], it is suspected that this situation may not be so severe for non-silicided cases.

For positive stress applied to V_{ss} with respect to V_{dd}, or negative stress applied to V_{dd} with respect to V_{ss}, the situation is different. The internal failure was again seen in an inverter, as shown in Figure 5.37(A). However, the damage does not extend to the p^+ drain of the pMOS device. This can be explained by considering the stress current path as shown in Figure 5.37(B). With positive stress on V_{ss}, the stress current path is through a lateral *npn*, formed with the n^+ source diffusion of the nMOS device, p^+ substrate, and the n^+ guardring of the pMOS device. As noted previously, this would be more severe for chips with silicided diffusions. In the case of devices with substrate bias generators, such as DRAMs, there is no effective low impedance current path to offer protection for this stress condition. To divert the stress current path, a protection diode may be employed between V_{dd} and V_{ss} as shown in Figure 5.38. Such an approach would be effective only for positive stress on V_{ss} with respect to V_{dd}. This type of protection could be distributed throughout

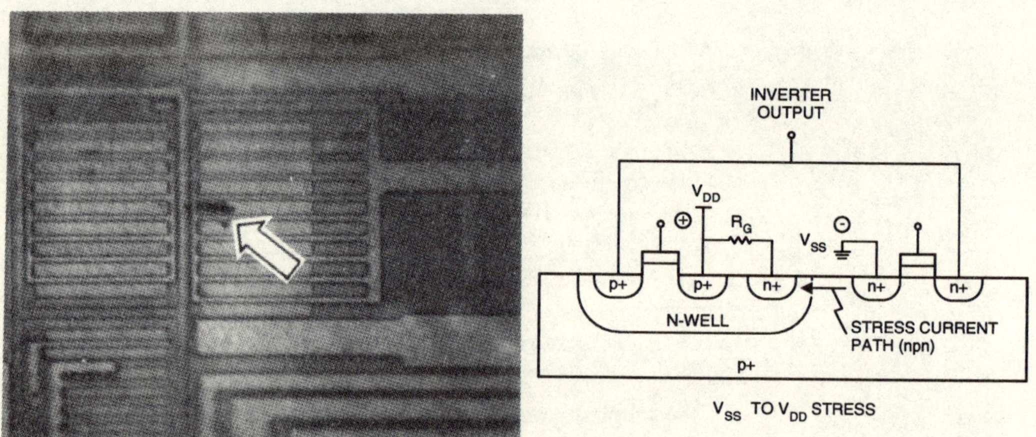

Figure 5.37 Damage site observed in an output buffer for negative stress on V_{dd} with respect to V_{ss}. The *p*-channel device is on the left side of the schematic.

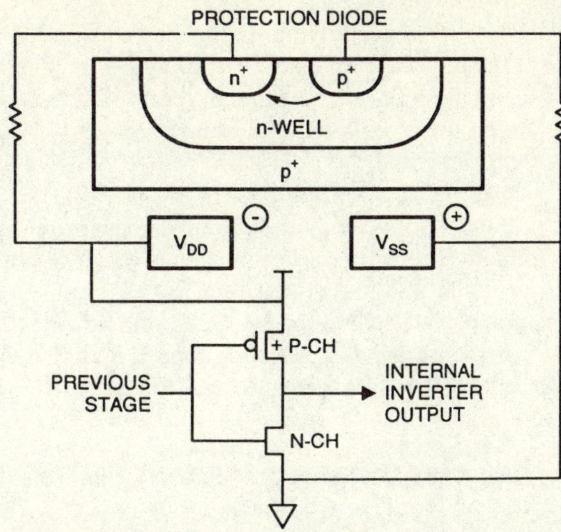

Figure 5.38 Schematic of a V_{ss} to V_{dd} protection using a diode to conduct in the forward mode for a positive voltage stress on V_{ss}.

the chip to attain overall protection for V_{ss} to V_{dd} stress. A chip with grounded substrate may not necessarily need this additional diode since the n^+ V_{dd} to p-substrate diode is intrinsic to the circuit. ICs fabricated in advanced CMOS processes employing the above protection techniques in the internal chip layout were found to be virtually free of the damage phenomena shown in Figures 5.36 and 5.37.

Internal chip layouts for improving circuit performance can also cause unexpected ESD failures. Consider the internal circuit schematic shown in Figure 5.39. The polysilicon-1 layer forms the gates of the internal transistors. The polysilicon-2 layer connected to V_{ss} is used as a means to decouple the noise on the gates of these

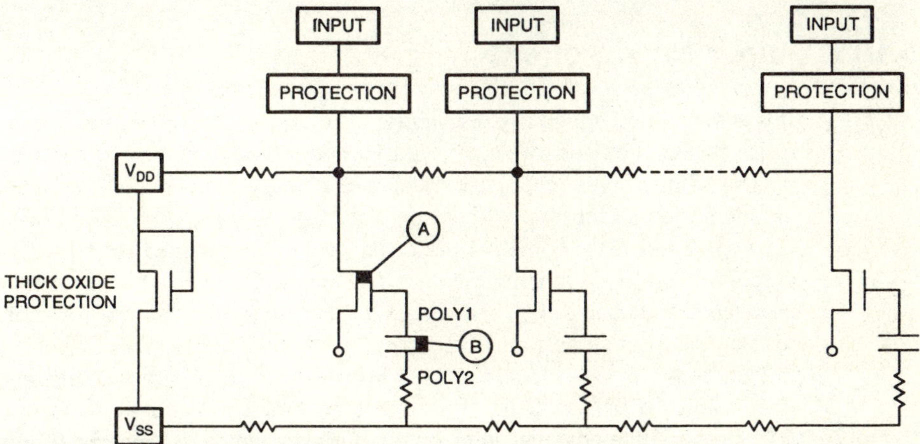

Figure 5.39 Internal circuit damage caused by positive stress between V_{dd} and V_{ss}.

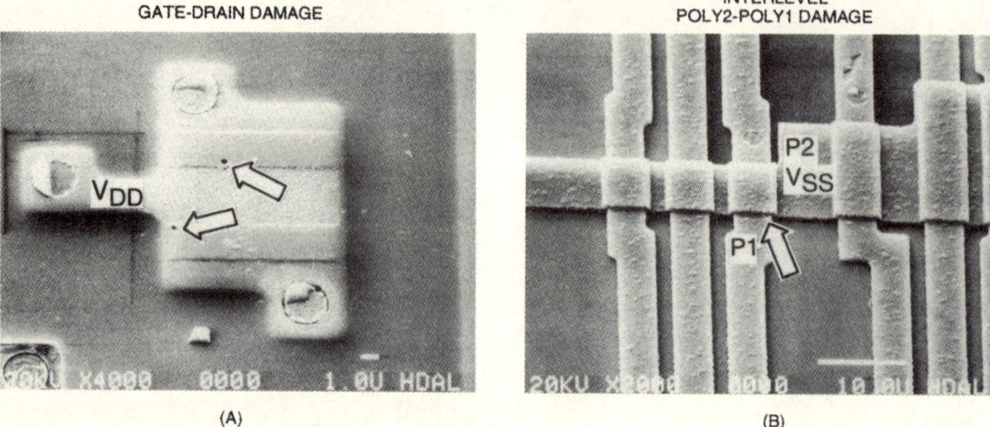

Figure 5.40 (A) Gate–drain damage at location A in Figure 5.39. (B) Interlevel oxide damage at location B in Figure 5.39.

transistors. However, this technique is found to have an adverse impact on the ESD performance. By testing the chip with positive ESD pulses between V_{dd} and V_{ss} it was found that failures occurred with only 2 kV of stress according to the Human Body Model test. Additionally, it was discovered that after V_{dd}–V_{ss} stress the input pins shown in Figure 5.39 were shorted to V_{ss}. Although a large thick field device existed between V_{dd} and V_{ss} (see Figure 5.39), it was found to be ineffective in this case. This was because the layout resulted in stress current path to ground. The internal damage occurred as a gate–drain short at location A and an interlevel oxide rupture at location B, as indicated in the figure. These fail modes are illustrated in Figures 5.40(A) and 5.40(B), respectively. An obvious modification that improves the ESD performance for V_{dd}–V_{ss} stress is to eliminate such stress current paths to ground by removing the noise decoupling polysilicon-2 layer at these locations.

5.10.2 Output to V_{dd} stress

It was mentioned previously that applying ESD stress between input pad and V_{dd} can cause current to flow between V_{dd} and V_{ss} and lead to damage in the internal parasitic devices. Similarly, internal parasitic device damage could also occur when a CMOS output buffer is stressed with respect to V_{dd}. It was reported by Duvvury *et al* [Duvvury87A] that good ESD performance can be obtained for the case of CMOS output buffers by optimizing the layout of the pMOS device to improve the *pn* diode between the pad and V_{dd} (as shown in Figure 4.49). The layout minimizes the resistance of the diode formed by the p^+ diffusion and *n*-well contact. However, in the limiting case for positive stress with respect to V_{dd} excessive current could circumvent the pMOS device path and take an alternative path as shown in Figure 5.41. This will trigger the lateral *npn* (formed with n^+ drain, *p*-substrate, and n^+ source) as well as other series *npn* devices between V_{ss} and V_{dd}, as indicated in the

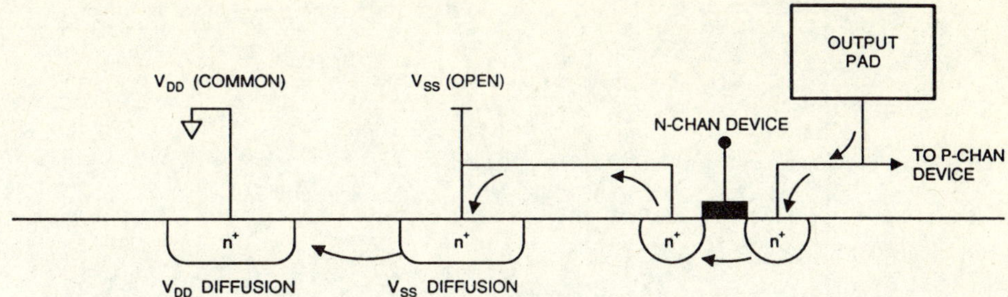

Figure 5.41   Circuit schematic of an output buffer showing current path during a $V_{dd}-V_{ss}$ stress.

figure. If an advanced CMOS process is used with silicided diffusions, the nMOS output device will be likely to show damage. Since complicated $V_{dd}-V_{ss}$ current paths are involved, the damage may also occur in one of the many parasitic *npn* devices in the internal circuitry. Therefore a careful consideration of this phenomenon must also be made in optimizing the internal chip layout.

5.11 STRESS DEPENDENT ESD BEHAVIOR

We saw earlier that a window of ESD failures can occur for ineffective protection designs. This phenomenon could also occur due to internal current paths. In most qualification procedures for ESD performance, the chips are usually stressed at a single ESD voltage level given by the *specification* and if it passes this test it is considered to have qualified. This is known as a 'go/no-go' method. For example, if the chip passes a stress level of 4 kV (according to the Human Body Model), it is assumed to pass 2 kV also. On the other hand, if it fails a stress level of 2 kV, it is assumed that it will not pass 4 kV. It will be shown here that this may not necessarily be the case if there are interactions with internal parasitic devices.

In Figure 5.42 the overall ESD performance of a VLSI chip is shown as a function of ESD stress level. These devices were stressed according to the Machine Model with all pins positive to V_{ss} and negative to V_{dd}. On the y-axis the percentage number of devices passing full functional test after ESD stress is shown and the x-axis gives the applied ESD stress level. There is clearly a window where the ESD performance of this device is weak. In this particular case, the failing devices are found to have a drastic increase in the V_{dd} to V_{ss} current, I_{dd}, after ESD stress. A more thorough investigation led to the conclusion that the I_{dd} current increase in the failed devices occurred only when the stress applied was negative with respect to V_{dd}. After more detailed data the particular I/O pins that are causing this phenomenon were isolated.

Although the results shown in Figure 5.43 are for the Human Body Model stress, it is still seen that when these I/O pins are stressed negative with respect to V_{dd}, an I_{dd} failure window, similar to the one in Figure 5.42, does exist. Liquid crystal analysis in the case of failed devices revealed the damage sites to be in the internal clock buffers.

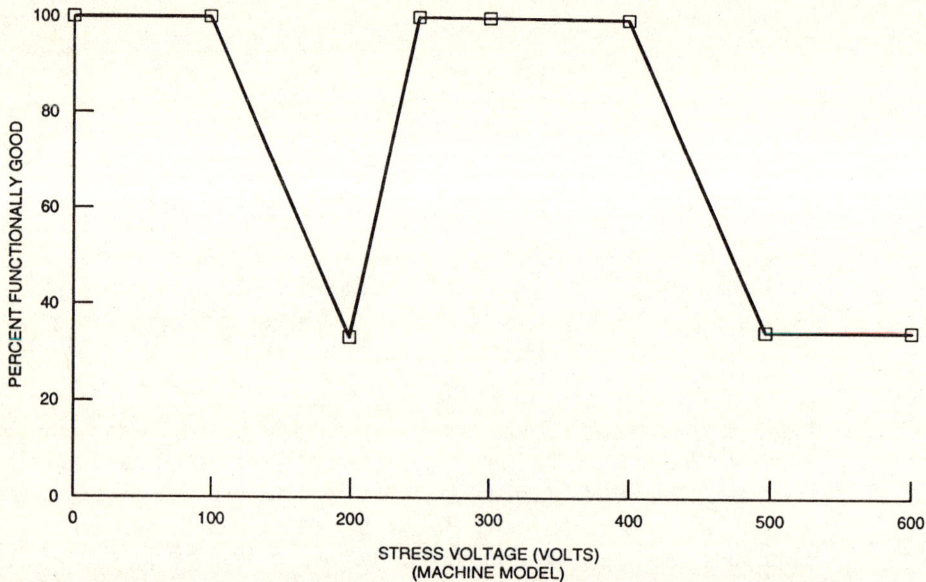

Figure 5.42 Functionally good device post-ESD stress vs. stress voltage.

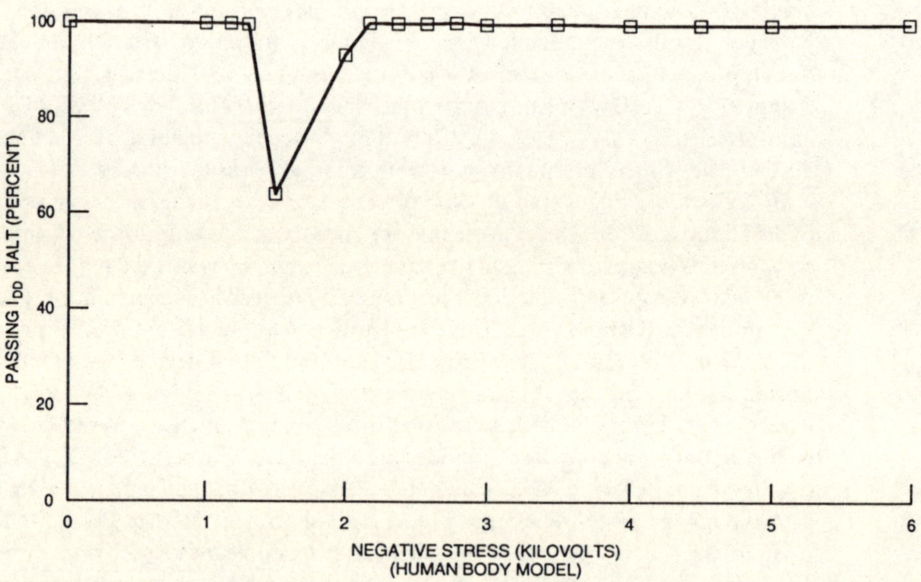

Figure 5.43 Devices passing I_{dd} leakage test post-ESD stress vs. stress voltage. Applied stress was negative with respect to V_{dd}.

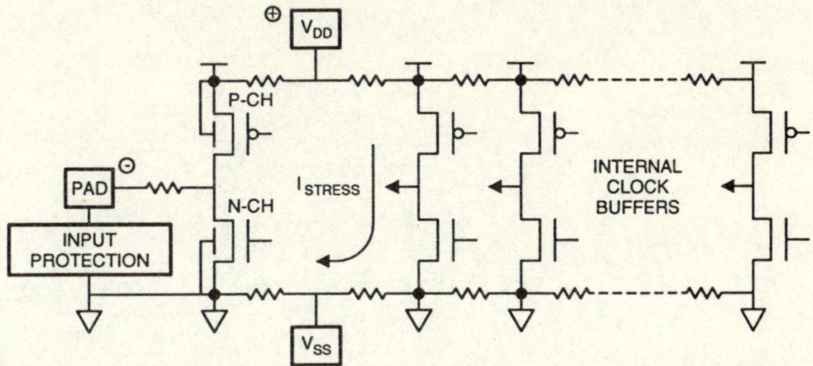

Figure 5.44 Layout schematic showing current path for negative stress between an I/O pin and V_{dd}.

The damaged clock buffer, as shown in Figure 5.44, is CMOS. With a negative stress to V_{dd}, the parasitic n^+ drain to substrate diode in the protection circuit becomes forward-biased and the stress current has to flow between V_{dd} and V_{ss}, as indicated in the figure. At relatively low stress levels (1.5 kV for the Human Body Model or 150 V for the Machine Model) the stress current caused damage in the parasitic *pnpn* (SCR) devices in the clock buffers. The stress current path responsible for this damage is highly layout dependent as shown in Figure 5.45. Note that the V_{dd}–V_{ss} protection circuit indicated in the figure could not have prevented the failure. The damage was identified through liquid crystal analysis as an nMOS gate–drain short in one or two of the clock buffers as indicated in Figure 5.46. This short

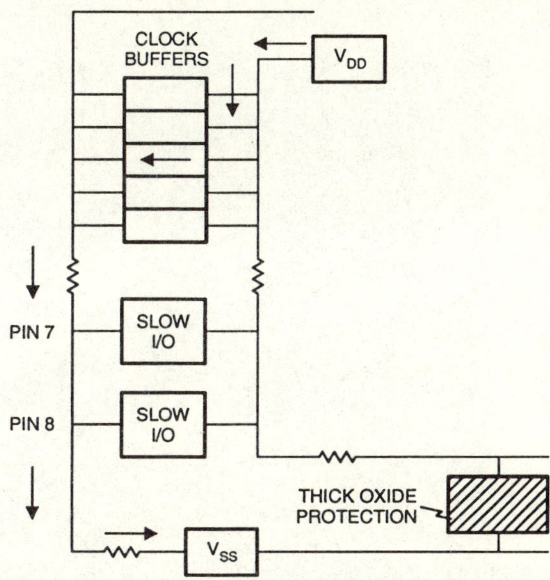

Figure 5.45 Block diagram of ESD protection, I/O buffers and internal clocks.

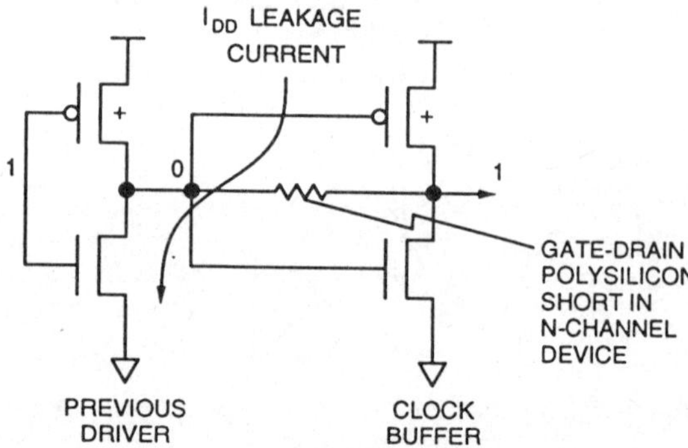

Figure 5.46 Damage in internal clock buffer with gate–drain short.

apparently caused self-biasing of the damaged buffer and manifested itself as increased I_{dd} current (see Figure 5.47). It is hypothesized that at lower stress levels the parasitic SCR in only one of the clock buffers was triggered. Therefore, this acted as a parasitic ESD protection circuit with a protection level of 1.5–2 kV. At higher stress levels it is likely that all five of the clock buffer SCRs were triggered. This would explain why no I_{dd} current increase or any damage in the clock buffers was observed at a stress of >2 kV. In fact, with all five SCRs triggered no I_{dd} failures would be observed up to >6 kV as shown in Figure 5.43. However, at these high stress levels the ESD protection circuits begin to fail and the chip functionality is affected as shown in Figure 5.42 for the Machine Model. One possible solution to

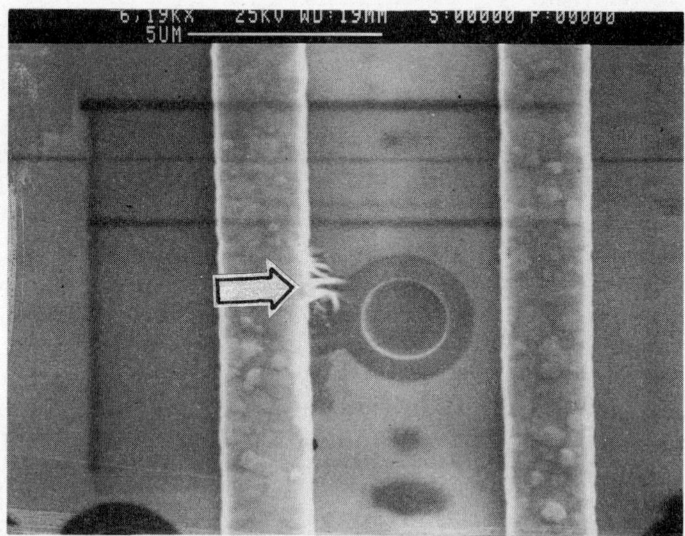

Figure 5.47 Circuit schematic of clock buffer showing I_{dd} leakage current path.

eliminate this failure would be to increase the individual clock buffer sizes to improve the parasitic SCR protection.

The type of stress dependent ESD behavior discussed above is unusual but not uncommon. Effectively designed input protection circuits tend to perform very well at 4 or 6 kV but often cause pin leakage at stress levels of 2 kV or below. Failure windows of this type point out the importance of step-stressing the device pins to establish the overall ESD immunity of a circuit chip.

The $V_{dd}-V_{ss}$ protection design requirements and how these can lead to internal chip ESD damage phenomena were discussed here. Implementing thick-field oxide devices between the V_{dd} and V_{ss} power bus lines could offer protection against stress between the two but only as long as the weak internal parasitic devices are eliminated. An example was shown where n^+ diffusions connected to V_{dd} and V_{ss} respectively should not be too close to each other in the internal chip layout.

5.12 SUMMARY

Failure analysis techniques have been discussed in this chapter together with some of the principal failure modes and their associated electrical signatures have been presented. Liquid crystal analysis and photon emission microscopy are extremely useful tools for the identification of failure locations. In addition, photon emission is an effective tool for the analysis of the operation of protection elements under high current conditions and in the debugging of weak designs and layouts of protection circuits. Failure modes can extend from *soft* to *severe* with post-stress leakage currents ranging from 1 nA to 1 mA and more in some cases. The most common failure mode in a properly designed circuit was that due to silicon melting as a result of current localization. In advanced CMOS circuits it was also shown that damage at the diffusion edge can be the dominant failure mechanism with leakage currents between 1 nA and 10 μA.

The effects of ESD stress between various pin combinations (such as output to V_{dd}) can cause ESD stress current to flow between V_{dd} and V_{ss} and cause internal ESD damage. It has been shown in this chapter that careful analysis of these issues is needed and proper layout techniques must be employed to ensure the good overall ESD performance in an IC.

Results from the analysis of a specific chip were reported. These show that when specific I/O pins are stressed negative with respect to V_{dd}, I_{dd} increases after a low level ESD stress (1.5 kV) but not after a higher stress level of 4 kV. This was explained by the damage caused by the stress current through the large internal CMOS clock buffers on the chip which caused them to act as parasitic SCR devices. For lower stress levels only one clock buffer SCR was triggered but at higher stress levels all five of the clock buffers triggered to provide high levels of ESD protection. It was pointed out that the possible existence of such stress dependent ESD performance should be determined by step-stressing the devices and testing for functionality.

The actual case studies presented in this chapter clearly illustrated the importance of internal ESD phenomena. In the current advanced technology VLSI chips and in the next generation of very high density ICs, the protection might very well be dictated by internal $V_{dd}-V_{ss}$ stress currents either as a result of direct stress between the two or as a result of indirect stress such as when I/O pins are stressed with respect to V_{dd} [Merrill93]. A good protection scheme would, of course, overcome this problem as well as provide robust protection circuits at the pins. Aside from the obviously visible internal ESD damage, the $V_{dd}-V_{ss}$ stress currents might possibly cause latent failures in the internal circuitry. Other reliability stress analyses are necessary to thoroughly understand these effects. For example, it has been shown that ESD stress can increase the susceptibility of nMOS devices to hot carrier degradation [Aur88]. The significance of such issues in the next generation of ICs will need more detailed study especially with regard to latent ESD effects.

REFERENCES

[Amerasekera90] A. Amerasekera, L. J. van Roozendaal, J. Abderhalden, J. J. P. Bruines, L. Sevat, 'An Analysis of Low Voltage ESD Damage in Advanced CMOS Processes', in *Proc. 12th EOS/ESD Symposium*, p. 143–150, 1990.

[Amerasekera92] A. Amerasekera, W. van den Abeelen, L. J. van Roozendaal, M. Hannemann, P. J. Schofield, 'ESD Failure Modes: Characteristics, Mechanisms and Process Influences', *IEEE Trans. Elec. Dev.*, ED-39, p. 2, 1992.

[Aur88] S. Aur, A. Chatterjee, T. Polgreen, 'Hot Electron Reliability and ESD Latent Damage', *IEEE Trans. Elec. Dev.*, ED-35, p. 2189–2193, 1988.

[Cavone94] M. Cavone, M. Nuschietiello, R. Rivoir, M. Stucchi, 'A Method for Characterization and Evaluation of ESD Protection Structures and Networks', in *Proc. 16th EOS/ESD Symposium*, p. 292–300, 1994.

[Colvin93] J. Colvin, 'The Identification and Analysis of Latent ESD Damage on CMOS Input Gates', in *Proc. 15th EOS/ESD Symposium*, p. 109–116, 1993.

[Cook93] C. Cook, S. Daniel, 'Characterization of New Failure Mechanisms Arising from Power-Pin ESD Stressing', in *Proc. 15th EOS/ESD Symposium*, p. 149–156, 1993.

[Duvvury83] C. Duvvury, R. N. Rountree, L. S. White, 'A Summary of Most Effective Electrostatic Protection Circuits for MOS Memories and Their Observed Failure Modes', in *Proc. 5th EOS/ESD Symposium*, p. 181–184, 1983.

[Duvvury86] C. Duvvury, R. McPhee, D. Baglee, R. Rountree, 'ESD Protection Reliability in 1-μm CMOS Circuit Performance', in *Proc. 24th IRPS*, p. 199–208, 1986.

[Duvvury87A] C. Duvvury, R. N. Rountree, Y. Fong, R. A. McPhee, 'ESD Phenomena and Protection Issues in CMOS Output Buffers', in *Proc. 25th IRPS*, p. 174–180, 1987.

[Duvvury87B] C. Duvvury, R. N. Rountree, R. A. McPhee, 'ESD Protection: Design and Layout Issues for VLSI Circuits', *IEEE Trans. Ind. Appl.*, IA-25, p. 41–47, 1987.

[Duvvury88A] C. Duvvury, R. Rountree, O. Adams, 'Internal Chip ESD Phenomena beyond the Protection Circuit', *IEEE Trans. Elec. Dev.*, ED-35, p. 2133–2139, 1988.

[Duvvury88B]   C. Duvvury, R. Rountree, 'Output ESD Protection Techniques for Advanced CMOS Processes', in *Proc. 10ᵗʰ EOS/ESD Symposium*, p. 206–211, 1988.

[Duvvury92]   C. Duvvury, C. Diaz, T. Haddock, 'Achieving Uniform NMOS Power Distribution for Improving ESD/EOS Reliability', in *Tech. Dig. IEDM* p. 131–134, 1992.

[Feng86]   W.-S. Feng. T. Y. Chan, C. Hu, 'MOSFET Drain Breakdown Voltage', *Elec. Dev. Lett.*, EDL-7, p. 449–452, 1986.

[Hannemann90]   M. Hannemann, A. Amerasekera, 'Photon Emission as a Tool for ESD Failure Localization and as a Technique for Studying ESD Phenomena', in *Proc. 1ˢᵗ European Symp. on Rel. and Failure Anal.*, p. 77–84, 1990.

[Hsu82]   F. C. Hsu, P. K. Ko, S. Tam, C. Hu, R. Muller, 'An Analytical Breakdown Model for Short-Channel MOSFETs', *IEEE Trans. Elec. Dev.*, ED-29, p. 1735–1740, 1982.

[Hulett81]   T. V. Hulett, 'On-Chip Protection of NMOS Devices', in *Proc. 3ʳᵈ EOS/ESD Symposium*, p. 90–96, 1981.

[Keller81]   J. K. Keller, 'Protection of MOS Integrated Circuits from Destruction by Electrical Discharge', in *Proc. 3ʳᵈ EOS/ESD Symposium*, p. 73–79, 1981.

[Kiefer93]   S. Kiefer, R. Milburn, K. Rackley, 'EOS Induced Polysilicon Migration in VLSI Gate Arrays', in *Proc. 15ᵗʰ EOS/ESD Symposium*, p. 123–128, 1993.

[Krakauer94]   D. Krakauer, K. Mistry, 'Circuit Interactions during Electrostatic Discharge', in *Proc. 16ᵗʰ EOS/ESD Symposium*, 1994.

[Kuper93]   F. Kuper, J. Bruines, J. Luchies, 'Suppression of Soft Failures in a Submicron CMOS Process', in *Proc. 15ᵗʰ EOS/ESD Symposium*, p. 117–122, 1993.

[Maene92]   N. Maene, J. Vandenbroeck, L. van den Bempt, 'On Electrostatic Discharge Protections for Inputs, Outputs and Supplies of CMOS Circuits', in *Proc. 14ᵗʰ EOS/ESD Symposium*, p. 228–233, 1992.

[Maloney88]   T. J. Maloney, 'Designing MOS Inputs and Outputs to Avoid Oxide Failure in the Charged Device Model', in *Proc. 10ᵗʰ EOS/ESD Symposium*, p. 220–227, 1988.

[McPhee86]   R. McPhee, C. Duvvury, R. N. Rountree, 'Thick Oxide Device ESD Performance under Process Variations', in *Proc. 8ᵗʰ EOS/ESD Symposium*, p. 173–181, 1986.

[Merrill93]   R. Merrill, E. Issaq, 'ESD Design Methodology', in *Proc. 15ᵗʰ EOS/ESD Symposium*, p. 233–237, 1993.

[Palella85]   A. Palella, H. Domingos, 'A Design Methodology for ESD Protection Networks', in *Proc. 7ᵗʰ EOS/ESD Symposium*, p. 169–174, 1985.

[Rountree85]   R. N. Rountree, C. L. Hutchins, 'NMOS Protection Circuitry', *IEEE Trans. Elec. Dev.*, ED-32, p. 910–917, 1985.

[Rountree88]   R. N. Rountree, C. Duvvury, T. Maki, H. Stiegler, 'A Process Tolerant Input Protection Circuit for Advanced CMOS Processes', in *Proc. 10ᵗʰ EOS/ESD Symposium*, p. 201–205, 1988.

[Wills88]   S. Wills, C. Duvvury, O. Adams, 'Photon Emission Testing for ESD Failures, Advantages and Limitations', in *Proc. 10ᵗʰ EOS/ESD Symposium*, p. 53–61, 1988.

6

MODELING AND CHARACTERIZATION

6.1 INTRODUCTION

There is no universal technique for designing ESD protection circuits. Through trial and error, designers have realized that protection circuits which work extremely well in a particular technology or IC can fail drastically when used in a different application or placed in a new technology. Since there are a number of factors that influence the behavior of a protection circuit, each of which may change in a new or different technology, the non-portability of ESD protection circuits comes as no surprise. Hence, the development of protection circuits for new technologies, or the use of existing protection circuits for new applications, requires a number of design iterations and testing. Experienced ESD protection circuit designers will require at least three iterations before the specified ESD level can be guaranteed. Hence, the cycle times for the characterization of the ESD robustness of a technology can exceed the total available time for the introduction of a new technology. Furthermore, the need to experimentally optimize the ESD performance of a protection circuit in a specific chip increases the product development cycle time.

Over the years, IC manufacturers have begun to integrate ESD characterization into the early development phase of new technologies [Daniel90][Krakauer92] [Amerasekera94A]. At the same time, IC designers have considered the ESD protection circuit requirement in parallel with input and output circuit design and area allocations [Duvvury88][Polgreen89]. However, the need to use computer simulations to model the behavior of the ESD protection structures and their performance in a circuit is becoming more apparent. Even simple modeling allows technology and circuit designers to rapidly evaluate the ESD robustness of their designs, and to reduce the number of iterations required to obtain good protection circuits [Chatterjee91]. Another major attraction is that modeling also enables ESD robustness to be designed into a new technology [Amerasekera93A], thereby ensuring that circuit designers have a solid foundation upon which to develop good ESD protection circuits. By rapidly exploring the technology design space, simulation techniques can enable the technology designer to optimize for ESD robustness as well as device performance.

In this chapter, we will discuss the basis for modeling ESD in basic protection circuit elements, and the approaches being taken. ESD failure is the result of thermal damage in the semiconductor and most ESD modeling techniques are based on an analysis of the heat dissipated in the semiconductor element during a high current stress event [Krieger87][Pierce88][Lin90][Dwyer90][Amerasekera91] [Diaz93A][Diaz93B]. More detailed models have coupled the electrical behavior of the elements with the thermal behavior in electrothermal models. A simple approach is to use lumped circuit elements to describe the thermal resistances and capacitances and implement the full model in a circuit simulator [Scott86][Beltman90] [Kurimoto94]. A more complex technique is to couple the full heat equation with the device equations in a complete electrothermal model of the element. Such models have been used for some time in power bipolar analysis and thyristor development [Gaur76][Adler78]. The more recent advances in computing have enabled these models to be applied to ESD type high currents in semiconductor circuit elements [Krabbenborg91][Mayaram91][Amerasekera93B].

The simulation of circuit behavior during ESD events is a much more complicated issue. There are a number of problems associated with obtaining a fully-coupled electrothermal circuit analysis, the biggest of which is the available computing power. However, approximations have been made which have enabled some analysis of the circuit behavior to be made [Chatterjee91][Duvvury92A] [Diaz93B]. The simplest approach to date is the use of circuit simulation tools such as SPICE with the maximum current and voltage of the individual circuit elements used as the boundary conditions. Advances in mixed-mode simulators which couple device modeling with circuit modeling have shown that it is possible to estimate the operating thresholds for ESD protection circuits [Duvvury92A][Diaz93B].

6.2 THE PHYSICS OF ESD DAMAGE

The high current behavior of semiconductor elements used in ESD protection circuits has been presented in Chapter 3. The circuits which are most commonly used are the lateral *npn* in MOS technologies (Figure 6.1(A)), and the vertical *npn* in bipolar and BiCMOS technologies (Figure 6.1(B)). Both these elements operate in a two-terminal mode during an ESD stress, relying on the avalanche generation of holes to provide the base current required to turn them on. Once on, the *npn* conducts current between the collector and the emitter with an on-resistance of a few ohms, and a holding voltage of between 5 V and 10 V depending on the technology. Most of the holding voltage is dropped across the collector-base junction and is required to maintain the avalanche generation of holes which provides the *extrinsic* base current to maintain the *npn* in the *on* state. Hence, almost all the power in the high current operating mode can be assumed to be generated in the collector-base junction. It follows that a lower collector-base voltage will result in a lower power dissipation in the device and will result in a higher current capability.

Note: The extrinsic base current is the current which flows in the extrinsic base

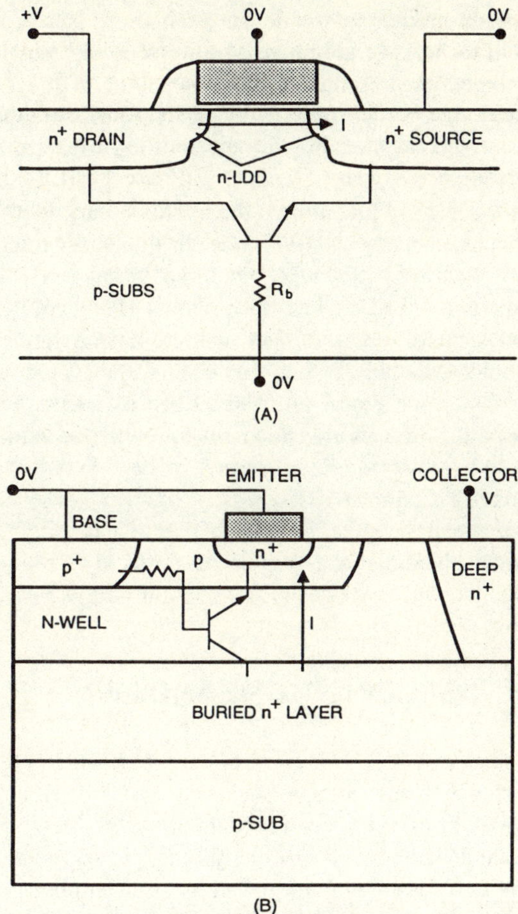

Figure 6.1   (A) Cross-section of an nMOS transistor showing the lateral *npn* transistor and the current flow, *I*. (B) Cross-section of a vertical bipolar *npn* transistor showing the direction of current flow, *I*.

resistance R_b as opposed to the intrinsic base current which flows across the forward-biased emitter base junction. This difference is important in understanding the self-biasing mechanism of lateral npn transistor action.

When a device is subjected to high current injection, the power dissipated in the device results in an increase in the internal temperature. At some point, given that there is no electrothermal effect on the device parameters, the internal temperature will approach that of the melting point of the semiconductor material (1685 K for silicon). A change of phase will then take place and the device properties will be irreversibly altered. However, it is well known that thermal breakdown occurs at temperatures below the melting point [Tauc57][Scarlett63][Melchior64]. This breakdown has been termed *second breakdown*, distinguishing it from the avalanche breakdown which occurs at lower injection currents. Second breakdown results in a collapse in the voltage across the device, and is significant in that the elec-trical properties of the device such as off-state current and the I–V curves are ob-served to be severely changed after second breakdown. Therefore, the onset of second breakdown is the damage threshold of the device defined by the second breakdown trigger current, I_{t2}. Since second breakdown is thermally initiated, models for second breakdown focus on the temperature rise in the device due to the power dissipated.

Once the second breakdown threshold is reached, the device goes into a negative differential resistance (NDR) mode. In the NDR mode, the voltage and current are unstable [Ridley63][Shaw92], and the voltage will decrease until a stable bias condition is reached. It has been shown [Ridley63][Khurana66][Shaw92] that the stable bias condition is in fact a microplasma formed by the constriction of the current into a filament. When this occurs, the current density in the filament is very high, and the melt temperature of silicon is reached rapidly in the localized region. The formation of a molten filament following second breakdown has been used to explain the failure modes observed in devices which undergo second breakdown [Khurana66][Nienhuis66][Brown72][Smith73][Amerasekera92] including junction edge damage and contact burn-out. By understanding the mechanisms of the thermal breakdown phenomena, it should be possible to model the ESD behavior of these structures as well as to design more robust ESD structures.

6.3 THERMAL ('SECOND') BREAKDOWN

The conditions for second breakdown are defined by the internal temperature of the device at which the voltage begins to collapse. However, the mechanisms responsible have not as yet been well defined in the literature. Limitation in the available experimental techniques in determining the temperature rise due to a given input power is one reason for the difficulties in defining the precise mechanisms governing the onset of second breakdown. The other primary limitation is the difficulty in numerically modeling the processes leading to second breakdown. There has been some success with the second problem in recent years with the advent of electro-thermal simulators; however, it is this problem that has been the reason for the main approximations used for the critical temperature at second breakdown.

Tauc and Abraham [Tauc57] first suggested that thermal generation of carriers could be responsible for permanent damage seen in germanium *pn* junctions operated in the breakdown region. They also observed that the decrease in voltage caused by thermal breakdown was accompanied by a current constriction. Later Schafft and French [Schafft62][Schafft66] showed that the high current breakdown could be related to thermal effects based on their experiments on the relationship between the energy required for breakdown and the time to breakdown. The first criterion for thermal breakdown can be attributed to Melchior and Strutt [Melchior64] who concluded on the basis of experiments on *npn* transistors that the breakdown was triggered when the collector-base junction reached its intrinsic temperature, T_i. T_i was defined as the temperature at which the intrinsic carrier concentration, n_i, was equal to the background doping concentration given by N_d for *n*-type material. However, later experimental work by Chen, Portnoy and Ferry [Chen71] and simulations by Ward [Ward76] and Orvis *et al* [Orvis83] were not able to confirm, or refute, that thermal breakdown was definitely initiated at T_i. In the absence of a suitable alternative, T_i has been used as a fitting parameter to obtain approximate solutions for the power required for thermal breakdown [Popescu70][Smith73][Ghandhi77][Alexander78].

In order to understand the concept of T_i and its relationship to thermal breakdown in more depth, we should consider the variation of the silicon resistivity with temperature for different doping concentrations as shown in Figure 6.2 [Runyan65]. As the temperature increases, the resistivity first increases as the mobility decreases. At a critical temperature given by T_c, the resistivity reaches a maximum and then decreases with increasing temperature. The maximum occurs when the effect of the increase in n_i on the resistivity begins to dominate over the decrease in mobility. The conductance for an *n*-type material with a doping concentration N_d is given by

$$\sigma = N_d \times q \times \mu \tag{6.1}$$

and n_i is given by

$$n_i = A \cdot T^{\frac{3}{2}} \cdot \exp\left(-\frac{E_g}{2kT}\right) \tag{6.2}$$

The temperature dependence of mobility is theoretically determined from phonon scattering which gives [Bardeen50]

$$\mu \sim T^{-\frac{3}{2}} \tag{6.3}$$

which has been experimentally shown to hold for higher impurity concentrations $N_d \approx 2 \times 10^{17}/\text{cm}^3$ [Jacoboni77][Sze81]. At lower doping concentrations other

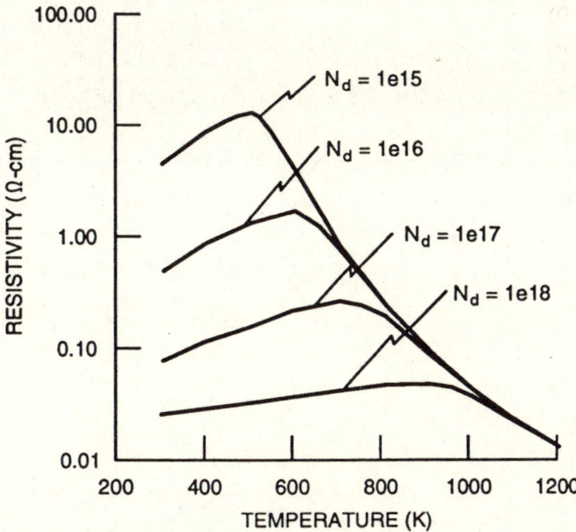

Figure 6.2 Resistivity as a function of temperature for different doping concentrations, N_d/cm^3.

scattering mechanisms result in an increased temperature dependence which has been determined empirically for n-type silicon [Jacoboni77]:

$$\mu \sim T^{-2.4} \tag{6.4}$$

The transition from a positive temperature coefficient of resistance to a negative temperature coefficient occurs when $n_i = N_d$. For higher temperatures, $n_i > N_d$, and N_d in Equation (6.1) is replaced by the temperature dependent n_i. Hence, for a purely resistive material with no other carrier generation or injection mechanisms $T_c = T_i$ can be taken to be the temperature at which the I–V curve shows NDR characteristics.

In reverse-biased semiconductor junctions there are additional mechanisms which add complexity to the analysis described above. It has been argued [Smith73][Alexander78] that since the potential barrier at the junction is a function of n_i, as n_i increases the potential barrier decreases. When $n_i = n = p$, the potential barrier is reduced to zero and the device is in thermal breakdown. The basis for this argument is as follows. The quasi-fermi levels are given by ϕ_n and ϕ_p for electrons and holes respectively. For a pn junction, the pn product in the depletion layer is

$$pn = n_i^2 \exp\left(\frac{q(\phi_p - \phi_n)}{kT}\right) \tag{6.5}$$

The potential barrier is given by $(\phi_p - \phi_n)$ which is less than zero for a reverse-biased junction. When the potential barrier vanishes, then $\phi_p = \phi_n$, and $pn = n_i^2$, giving $p = n = n_i$. Again the assumption is that all generation is thermal, and that there is no avalanching taking place which makes this analysis only applicable at low reverse-biased voltages well below that required to cause avalanche breakdown. At high current levels, the junction is in avalanche breakdown, and simulations have shown that the avalanche generated carriers result in n_i^2 actually being much greater than pn at the second breakdown threshold [Ward77][Yee82][Orvis83] [Mayaram91][Amerasekera93B]. Thus the use of $T = T_i$ as the condition for second breakdown can only be regarded as an approximation for ESD type second breakdown.

The conditions for second breakdown can be more generally defined by considering the terminal current and voltage across a device [Burgess60] [Shaw92][Amerasekera93B]. The $I - V$ curve for a typical reverse-biased pn junction is shown in Figure 6.3. At the second breakdown point, the voltage reaches a maximum and as the current increases further, the voltage decreases. In general the current is given by

$$I = I(V, T) \tag{6.6}$$

and, therefore,

$$\frac{\mathrm{d}I}{\mathrm{d}V} = \left.\frac{\partial I}{\partial V}\right|_T + \left.\frac{\partial I}{\partial T}\right|_V \cdot \frac{\mathrm{d}T}{\mathrm{d}V} \tag{6.7}$$

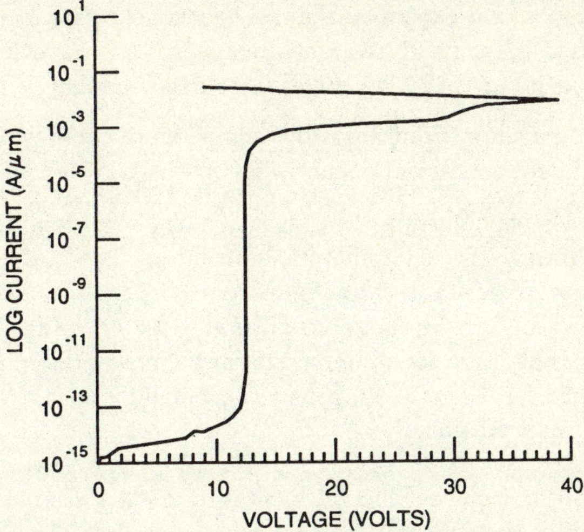

Figure 6.3 I–V curve for a *pn* diode in reverse bias.

Now the temperature can be determined from the heating in the device, and so $T = T(P)$, where $P = I \times V$. Hence,

$$\left.\frac{\partial T}{\partial V}\right|_I = I \cdot \frac{\mathrm{d}T}{\mathrm{d}P} \tag{6.8}$$

and

$$\left.\frac{\partial T}{\partial I}\right|_V = V \cdot \frac{\mathrm{d}T}{\mathrm{d}P} \tag{6.9}$$

Burgess [Burgess60], in an analysis of thermistor devices, showed that Equations (6.7)–(6.9) could be rearranged to give

$$\frac{\mathrm{d}V}{\mathrm{d}I} = \frac{V}{I} \cdot \frac{1-y}{x+y} \tag{6.10}$$

where

$$x = R \cdot \left.\frac{\partial I}{\partial V}\right|_T \tag{6.11}$$

and

$$y = V \cdot \frac{\mathrm{d}T}{\mathrm{d}P} \cdot \left.\frac{\partial I}{\partial T}\right|_V \tag{6.12}$$

At second breakdown, $\mathrm{d}V/\mathrm{d}I = 0$, and $y = 1$. The condition for second breakdown is then given by,

$$\left.\frac{\partial I}{\partial T}\right|_V \cdot \frac{\mathrm{d}T}{\mathrm{d}P} = 1 \tag{6.13}$$

Amerasekera *et al.* [Amerasekera93B] presented a similar condition for second breakdown, and explained the mechanism in terms of the electrothermal effects in the device. Essentially, second breakdown occurs when the heat produced by an increase in current, ΔI, generates the exact amount of minority carriers required to support the increase in current without raising the electric field. As the thermal heating increases, the heat generated by ΔI generates more carriers than are needed to support the increase in I, and the voltage decreases leading to the familiar *snapback* phenomenon. They showed that this condition was satisfied at second breakdown using electrothermal simulations of $p^+/n/n^+$ avalanche diodes.

However, the above condition is not easily applicable as a predictor, because of the non-linearities of the equations governing avalanche and thermal generation currents. It is possible to make assumptions for the power dissipation and the conductivity as described by Shaw *et al.* [Shaw92], but they do not really solve the problem of predictive modeling.

A first order model of the high current performance of an ESD protection circuit element can best be achieved by considering the maximum power that the element is capable of withstanding before damage occurs. This would enable the maximum current that can be passed through the element to be determined, which can subsequently be translated into an ESD failure threshold. The boundary condition used in the development of the model is the damage threshold under high current injection. Recent works [Krieger87][Pierce88][Lin90] took this boundary condition to be the temperature at which the melting temperature of silicon, 1412°C, is reached or, in the case of aluminum contacts, the eutectic temperature for Al–Si interdiffusion (550°C). More recently still, Polgreen and Chatterjee [Polgreen89] and Amerasekera *et al.* [Amerasekera90] showed that device failure was related to second breakdown at the stressed junction. However, since there is no real way of determining the actual device temperature at the onset of failure (although a failed device has obviously seen > 1412°C), it is necessary to extract the dependence of the failure power, P_f, as a function of stress time for failure, t_f, using experimental data [Pierce88][Dwyer90][Amerasekera91]. While this approach enables the ESD sensitivity of some design and process related parameters to be determined, it is not directly suitable for predictive modeling of ESD capability of a technology or device structure.

6.4 ANALYTICAL MODELS USING THE HEAT EQUATION

An analytical approach to ESD modeling starts with the solution of the heat conduction equation for the device geometry under consideration:

$$\frac{\partial T}{\partial t} - D \cdot \nabla^2(T) = \frac{q(t)}{\rho \cdot C_p} \tag{6.14}$$

where ρ is the density of the semiconductor in g/cm^3, C_p is the specific heat in J/g-K, $D = K/(\rho C_p)$ is the thermal diffusivity in cm^2/s, and K is the thermal conductivity in W/cm-K. $q(t)$ is the rate of heating per unit volume of the heat source.

The standard solution to the heat equation for a variety of sources and boundary conditions has been well documented by Carslaw and Jaeger [Carslaw59]. In modeling electrical overstress and ESD, Wunsch and Bell first presented a simple thermal model for the power as a function of the time [Wunsch68]:

$$\frac{P}{A} = \sqrt{\pi K \rho C_p} \cdot (T - T_0) \cdot t^{-\frac{1}{2}} \tag{6.15}$$

where P is the input power and A is the area of the heat source which is effectively the junction area. T is the temperature in the device and T_0 is the initial temperature. The great advantage of the equation is its simplicity and ease of use. The power dependence of time easily translates to a straight line on a log–log plot, and the coefficients can be lumped to give

$$P = B \cdot t^{-\frac{1}{2}} \tag{6.16}$$

where

$$B = A\sqrt{\pi K \rho C_p}(T - T_0) \tag{6.17}$$

is obtained from a set of experimental results and then used to determine the effect of a given transient stress on the failure threshold. The simplicity of the equation has ensured its wide use in the analysis of electrical overstress and ESD failure data [Speakman74][Mathews80][Pierce88][Diaz92].

A more detailed analysis of the heat equation leads to a solution of the form described by Tasca, Peden and Miletta [Tasca70][Tasca72]:

$$P = \left(\frac{A}{t} + B \cdot t^{-\frac{1}{2}} + C\right)(T - T_0) \tag{6.18}$$

The above equation shows that the time dependence of the power dissipation is determined by the time period of the applied pulse. For very short duration pulses, there is very little heat flow out of the defect region, and the time dependence follows an adiabatic or $1/t$ dependence. At longer stress times, the Wunsch–Bell equation is followed, and the third term in the parentheses simply describes the steady-state condition. Dwyer, Franklin and Campbell [Dwyer90] developed the equation further to show that there are actually four time-dependent regions. In addition to the three described by Tasca, they also showed that in the period immediately before steady state was reached, the dependence showed a $1/\ln(t)$ form. The four regions are shown in the curve in Figure 6.4, based on Dwyer's work.

The thermal diffusion length, $L = \sqrt{Dt}$, is of the order of 10 μm for a pulse duration of 1 μs. This is much smaller than the distance between the protection circuit and the edge of the chip or wafer. We can assume that the heat is being radiated into an infinite medium, since as either x or y tends to infinity, T tends to ambient. Similarly in the z direction, for $z > 0$ (vertical depth) the silicon thickness is much greater than the diffusion length, and we may assume an infinite medium for $z > 0$. Therefore, the problem can be reduced to one of a heat source of a given geometry in a semi-infinite medium $-\infty < x < \infty, -\infty < y < \infty$ and $z > 0$. Usually, the region $z < 0$ is covered by oxide and may be assumed to be a poor heat

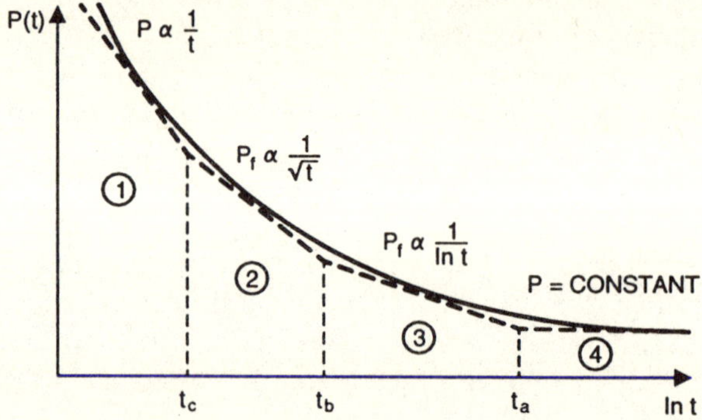

Figure 6.4 Power vs. time curve for a given temperature. As t is decreased the power required to reach that temperature increases. The analytical solutions to the heat equation show four regions in the curve.

conductor. The boundary $z = 0$ can, therefore, be considered to be reflective and, by the method of images, the heat source is mirrored in the region $z < 0$ [Carslaw59][Krieger87]. Krieger [Krieger87] has pointed out that in n-channel MOS structures, the direction $z < 0$ may not be thermally insulating because of the presence of the polycrystalline silicon gate and associated metalization which lies just above the junction close to the location of the heat source. This may have some effect on the analytical models described below, but the discrepancy is not expected to be significant.

The general solution for this problem can be found using the Green's function method [Carslaw59] whereby the solution to the equation

$$\frac{\partial G}{\partial t} - D \cdot \nabla^2(G) = \delta(r - r') \cdot \delta(t - t') \tag{6.19}$$

is given by

$$T(r, t, r', t') = \int_0^t dt' \int_r q(r', t') \cdot G(r, t, r', t') \, d^3r' \tag{6.20}$$

$G(r, t, r', t')$ is the Green's function for the given heat source. In three-dimensional rectangular coordinates the Green's function is

$$G(r, t, r', t') = \frac{1}{\{4\pi D(t - t')\}^{\frac{3}{2}}} \cdot \exp\left(-\frac{(r - r')^2}{4D(t - t')}\right) \tag{21}$$

where

$$(r - r')^2 = (x - x')^2 + (y - y')^2 + (z - z')^2 \tag{22}$$

For a rectangular heat source $-\frac{a}{2} < x < \frac{a}{2}, -\frac{b}{2} < y < \frac{b}{2}$, and $0 < z < \frac{c}{2}$;

$$G(x, a, t) = \frac{1}{2}\left[\text{erf}\left(\frac{\frac{a}{2} + x}{\sqrt{4Dt'}}\right) + \text{erf}\left(\frac{\frac{a}{2} - x}{\sqrt{4Dt'}}\right)\right] \tag{6.23}$$

$$G(y, b, t) = \frac{1}{2}\left[\text{erf}\left(\frac{\frac{b}{2} + y}{\sqrt{4Dt'}} \right) + \text{erf}\left(\frac{\frac{b}{2} - y}{\sqrt{4Dt'}} \right) \right]$$ (6.24)

$$G(z, c, t) = \frac{1}{2}\left[\text{erf}\left(\frac{z}{\sqrt{4Dt'}} \right) + \text{erf}\left(\frac{\frac{c}{2} - z}{\sqrt{4Dt'}} \right) \right]$$ (6.25)

The heat flow $q(t')$ is given by $2 \cdot P(t')/(a \cdot b \cdot c/2)$ where $P(t')$ is the input power in watts, and the factor of 2 accounts for the reflective boundary condition at $z = 0$.

Using the above equations, the temperature at a point r at time t is given by

$$T(r, t) = T_0 + \frac{P_0}{\rho C_p abc} \int_0^t G(x, a, t') \cdot G(y, b, t') \cdot G(z, c, t')\, dt'$$ (26)

where the power $P(t')$ is considered to be time-independent and equal to P_0.

Hence, the peak temperature at the center of the heat source $r = 0$ is

$$T(0, t) = T_0 + \frac{P_0}{\rho C_p abc} \int_0^t \text{erf}\left(\frac{a}{4\sqrt{Dt'}} \right) \cdot \text{erf}\left(\frac{b}{4\sqrt{Dt'}} \right) \cdot \text{erf}\left(\frac{c}{4\sqrt{Dt'}} \right) dt'$$ (6.27)

The diffusion times related to a, b, and c are defined as

$$t_a = \frac{a^2}{4\pi D}$$ (6.28)

$$t_b = \frac{b^2}{4\pi D}$$ (6.29)

$$t_c = \frac{c^2}{4\pi D}$$ (6.30)

and are the times required for the device to reach thermal equilibrium in the x, y and z directions of the device. If $a > b > c$, then after a time t_a, the device has reached its steady-state temperature and $dT/dt = 0$.

Dwyer approximated the error functions using

$$\text{erf}(x) \approx \begin{cases} \frac{2x}{\sqrt{\pi}} & \text{if } x \leq \frac{\sqrt{\pi}}{2} \\ 1 & \text{if } x \geqslant \frac{\sqrt{\pi}}{2} \end{cases}$$ (6.31)

to obtain an estimate of the value of the integral in Equation (6.27). The approximations in Equations (6.28) and (6.30) are excellent except in the range $0.22 < x < 1.82$, and are considered to be reasonable in this range. Using these approximations, the four time-dependent regions of the power vs. time curve in Figure 6.4 are given by

$$P(t) = \frac{\rho C_p abc \cdot (T - T_0)}{t}, \qquad 0 < t < t_c$$ (6.32)

$$P(t) = \frac{ab\sqrt{\pi K \rho C_p} \cdot (T - T_0)}{\sqrt{t} - \frac{\sqrt{t_c}}{2}}, \qquad t_c < t < t_b \qquad (6.33)$$

$$P(t) = \frac{4\pi K a \cdot (T - T_0)}{\ln\left(\frac{t}{t_b}\right) - 2 - \frac{c}{b}}, \qquad t_b < t < t_a \qquad (6.34)$$

$$P(t) = \frac{2\pi K a \cdot (T - T_0)}{\ln\left(\frac{a}{b}\right) + 2 - \frac{c}{2b} - \sqrt{\frac{t_a}{t}}}, \qquad t_a < t \qquad (6.35)$$

Equation (6.35) is the steady-state equation when t tends to infinity. Equation (6.33) reduces to the Wunsch–Bell equation given in Equation (6.15) for small t_c; i.e. when the heat source is effectively a two dimensional plane. Dwyer's analysis goes on to link the relationship between the different regions and the equations obtained by Wunsch and Bell [Wunsch68], Tasca [Tasca70] and Arkhipov *et al.* [Arkhipov83] in terms of the relative dimensions of a, b and c. If $a = b = c$, then the equations reduce to that of the adiabatic $(1/t)$ dependence and the steady-state condition in Equation (6.32) and Equation (6.35) respectively. The infinite cylinder described by Arkhipov *et al*, has $a = \infty$ and $b = c$, and the thermal behavior is described by Equations (6.32) and (6.34). Similarly other forms of heat sources can be approximated by combinations of Equations (6.32) to (6.35). Experimentally, the power vs. time curve can be used to obtain the actual dimensions of the heat source by determining the time at which the transition is made from one region to the next [Dwyer90].

While the above analysis allows the entire range of electrical overstress-related thermal failure in semiconductor structures to be studied, the ESD conditions actually require a much more limited approach. The ESD pulse duration is of the order of 100 to 200 ns, and can be studied using a constant current pulse with pulse widths in that range. The dimensions of the devices being stressed indicate that the time to failure typically lies in the interval $t_b < t < t_a$ [Dwyer90][Amerasekera91]. Therefore, one need only apply Equation (6.34) to the analysis of the behavior of ESD protection circuits. Using a combination of experimental and analytical methods, the process and design dimensions influencing ESD can be extracted and used to improve the ESD performance [Pierce88][Amerasekera91]. However, this is an interactive process performed during or after technology development in order to characterize the technology, and does not achieve the requirements of a predictive modeling technique.

6.5 ELECTROTHERMAL SIMULATIONS

Electrothermal simulations with full coupling between the electrical and thermal equations are essential in order to accurately describe the behavior of the device in the high current region close to second breakdown. In Section 6.3 it was shown that the onset of second breakdown was dependent on the rate of change of the avalanche and thermal generation currents with temperature. Hence, the coupling between the temperature and the current densities, impact ionization coefficients, mobilities and electric fields are important in simulating ESD phenomena in devices. Full coupling requires that the correct forms of the current flow equations in the presence of thermal gradients are used [Wachutka90][Selberherr84]. Following the method of Stratton [Stratton72], the current densities for electrons and holes, $\vec{J}_n$ and $\vec{J}_p$, are written as

$$\vec{J}_n = qn\mu_n \cdot \vec{E} + qD_n \cdot \vec{\nabla}n + qnD_n^T \cdot \vec{\nabla}T \qquad (6.36)$$

$$\vec{J}_p = qp\mu_p \cdot \vec{E} - qD_p \cdot \vec{\nabla}p - qpD_p^T \cdot \vec{\nabla}T \qquad (6.37)$$

where D_n and D_p are the diffusion constants for electrons and holes, and $D_n^T \approx D_n/2T$ and $D_p^T \approx D_p/2T$ are the thermal diffusion constants for electrons and holes [Stratton72], n and p are the electron and hole concentrations and μ_n and μ_p are the electron and hole mobilities. The thermal gradients denoted by $\vec{\nabla}T$ account for the additional driving force of temperature on the current which is the Seebeck effect [Geballe55].

These equations have been subsequently incorporated into rigorous treatments of heat generation and conduction in semiconductor devices [Alwin77] [Adler78][Chryssafis79][Wachutka90]. However, they increase the complexity of the simulations, resulting in longer computation times and increased difficulties in obtaining converged solutions. Therefore, many simulations for second breakdown [Gaur76][Ward76][Orvis83] used isothermal diffusion coefficients. Such simulations, while limited, improve the understanding of the phenomena involved in second breakdown by providing a qualitative insight into the phenomena involved. In particular the work of Ward [Ward76], Koyanagi, Hane and Suzuki [Koyanagi77] and Orvis *et al* [Orvis83] helped to show that second breakdown was indeed thermally initiated in the time durations being considered. The question of whether electrical or thermal effects were responsible for second breakdown had been asked for many years, based on work on snapback in avalanche diodes and *npn* transistors [Steele62][Grutchfield66][Roman70][Caruso74] [Hower70].

The heat source, H (W/cm^3), was originally considered simply from the Joule heating term $J \cdot E$ [Ward76]. However, the heat gained by the lattice through recombination was shown later to play an important part in the thermal process [Adler78]. More recently Wachutka [Wachutka90] presented a form of the heat source given by

$$H = \vec{J} \cdot \vec{E} + (R - G) \cdot (\mathscr{E}_g + 3kT) - \frac{\vec{J}}{q} \cdot \left(\frac{3}{2}\kappa(T) \cdot \vec{\nabla}T + \frac{1}{2}\nabla\mathscr{E}_g\right) \qquad (6.38)$$

where $\vec{J}$ is the total current density, $\vec{E}$ is the electric field, $\mathscr{E}_g$ is the energy gap, T is the temperature, R is the recombination rate, G is the generation rate, and $\kappa(T)$ is the temperature-dependent thermal conductivity. The first term on the right-hand side is the standard Joule heating term, the second term accounts for lattice heating due to recombination/generation, and the third term brings in the Thompson effect due to the heating (or cooling) that takes place when carriers traverse a region with spatially varying thermoelectric power, P_n or P_p [Callen60][Sze81], caused by large temperature gradients. In reverse-biased junctions Joule heating will be most dominant, with some influence of the Thompson heating in regions with large temperature gradients. In forward-biased junctions, Joule heating provides the initial temperature rise, while the recombination term plays a large role in the heating effect at higher temperatures [Adler78] [Krabbenborg91].

The temperature-dependent thermal conductivity $\kappa(T)$ is given by [Selberherr84]

$$\kappa(T) = \frac{1}{0.03 + 1.56 \times 10^{-3}T + 1.65 \times 10^{-6}T^2} \tag{6.39}$$

Mobility is modeled using the Caughey–Thomas empirical mobility model [Caughey67] for the concentration-dependent zero-bias mobilities μ_{0n} and μ_{0p} in cm^2/V-s for electrons and holes:

$$\mu_{0n,0p} = \mu_{n,p}^{min} + \frac{\mu_{n,p}^L - \mu_{n,p}^{min}}{1 + \left[\frac{(T/300)^{-3.8}}{N/N_{ref}}\right]^{\beta_{n,p}}} \tag{6.40}$$

where $\mu_{n,p}^L$ is the lattice mobility which is temperature dependent, $\mu_{n,p}^{min}$ is the coefficient for ionized impurity scattering, N is the local total impurity concentration and N_{ref} is a constant [Selberherr84]. $\beta_{n,p}$ is an empirically determined constant. The zero-bias mobility is then used to determine the mobility at high electric fields using the equation

$$\mu_{n,p} = \frac{\mu_{0n,0p}}{1 + \frac{\mu_{0n,0p} \cdot E}{v_{sat}}} \tag{6.41}$$

where v_{sat} is the temperature-dependent saturation velocity [Selberherr84],

$$v_{sat} = \frac{2.4 \times 10^7}{1 + 0.8 \cdot \exp\left(\frac{T}{600}\right)} \tag{6.42}$$

The impact ionization coefficients need to be temperature dependent, and there have been many empirical forms used in the literature [Chynoweth60][Crowell66] [Overstraeten70][Okuto75][Ward76]. The equation of Okuto and Crowell [Okuto75] gives an empirically determined form for the impact ionization coefficients for electrons and holes, $\alpha_{n,p}$/cm, as a function of temperature:

$$\alpha_{n,p} = A_{n,p} \cdot \{1 + C_{n,p} \cdot 10^{-4}(T - 300)\} \cdot E \cdot \exp\left(\frac{-B_{n,p}^2 \cdot [1 + D_{n,p} \cdot (T - 300)]^2}{E^2}\right) \tag{6.43}$$

where $A_n = 0.426$/V, $A_p = 0.243$/V, $B_n = 4.81 \times 10^5$ V/cm, $B_p = 6.53 \times 10^5$ V/cm, $C_n = 3.05 \times 10^{-4}$, $C_p = 5.35 \times 10^{-4}$, $D_n = 6.86 \times 10^{-4}$, and $D_p = 5.87 \times 10^{-5}$ are the coefficients for electrons and holes; E is in V/cm. The above equations can be simplified to obtain $\alpha_{n,p}$ by reducing to,

$$\alpha_{n,p} = C_1 \cdot \exp(-C_2/E) \tag{6.44}$$

where $C_2 = \mathscr{E}_g/q\lambda$ and λ is the mean free path of the carrier [Crowell66]:

$$\lambda = \lambda_0 \tanh(E_{r0}/2kT) \tag{6.45}$$

$\lambda_0 \approx 50$ Å and $E_{r0} = 50$ meV are λ and the optical phonon energy, E_r, at 0 K. This is the equation used in the electrothermal application module of TMA MEDICI [MEDICI92]. A reasonable fit to experimental results has been obtained by using [Grant73]

$$\alpha_{n,p} = C_1 \cdot \exp(-b(T)/E) \tag{6.46}$$

where the coefficient C_1 remains constant as a function of temperature and the major variation with temperature is assumed to occur in the exponent. $b(T)$ is assumed to have a linear dependence on temperature and for electrons $db_e/dT \approx 1.3 \times 10^3$ cm/V-K, while for holes $db_h/dT \approx 1.1 \times 10^3$ cm/V-K [Grant73]. The simplified forms can be used in the development of analytical models (e.g. [Abderhalden91]), but it is best to use the complete form in a full electrothermal model.

The temperature rise in the structure being simulated is a very strong function of the ambient conditions and, therefore, strongly dependent on the thermal boundary conditions [Amerasekera93B][Yang93]. The analytical models described in the previous section assumed that heat was dissipated into a semi-infinite medium. It was shown that, considering the dimensions of the silicon, these approximations are reasonable. However, in an electrothermal simulator, the implementation of a large volume of silicon surrounding the device leads to drastic increases in the computation time. Hence, it is prudent to use thermal boundary conditions closely resembling the actual conditions in the simulations.

The thermal boundary conditions at the contacts are implemented using a lumped thermal resistance, R_{th} K/W [Yang93]. The temperature at the contact, T_{cont}, is determined from

$$T_{cont} - T_{amb} = R_{th} \cdot \Lambda \tag{6.47}$$

Λ is the heat flux through the contact obtained from Equation (6.14). T_{amb} is the ambient temperature, and $T_{cont} = T_{amb}$ for $R_{th} = 0$ K/W. At the other boundaries, distributed thermal resistances should be used, allowing each boundary node to be contacted to a thermal resistance. In addition, to account for the heat capacity of the surrounding silicon, distributed thermal capacitances, C_{th}, need to be included at the boundaries. As would be expected, C_{th} plays a significant role in transient thermal simulations.

It should be noted that none of these models have been experimentally verified above temperatures of 1000 K. It is possible that at higher temperatures the behavior of these parameters will change. However, it has been observed that simulations of simple semiconductor structures using these models with internal device temperatures >1000 K show a reasonable agreement with the experimentally

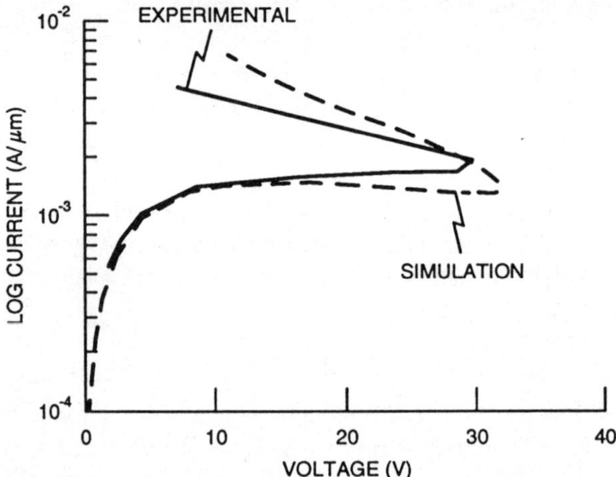

Figure 6.5 $I-V$ curve of an n-type resistor. Comparison between experimental result and that obtained from electrothermal simulation.

observed behavior as indicated in Figure 6.5 for an n-type resistor [Amerasekera-93B]. The temperature at which thermal snapback occurs is 1100 K for the simulated $I-V$ curve.

The $I-V$ curve of a $p^+/n/n^+$ diode is shown in Figure 6.6 as a function of the contact thermal resistance, R_{th}. Curve (1) has $R_{th} = 0$ K/W, Curve (2) has $R_{th} = 10^4$ K/W, and Curve (3) has $R_{th} = 10^6$ K/W. It is seen that R_{th} has a large influence on both the second breakdown voltage and the current. Such simulations have shown [Amerasekera93B] that the onset of second breakdown in diodes is the result of conductivity modulation taking place in the device. Conductivity modulation occurs

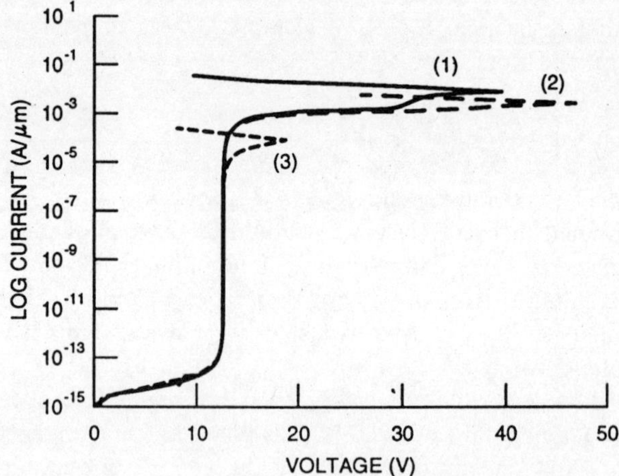

Figure 6.6 $I-V$ curves for a $p^+/n/n^+$ diode obtained from electrothermal simulations with different contact thermal resistors, R_{th}. Curve (1): $R_{th}=0$ K/W; Curve (2): $R_{th}=10^4$ K/W; Curve (3): $R_{th}=10^6$ K/W.

when the number of generated holes becomes large enough to support part of the current and, therefore, the electric field does not need to increase further to support an increase in the current. Typically, the onset of conductivity modulation occurs when $p \approx 0.2n$ [Amerasekera93B]. In the resistor structures, the simulations showed that the decrease in resistivity with increased temperature alone is not responsible for the collapse in voltage. It is essential that there is significant hole generation to enable the conductivity modulation mechanism to be initiated for second breakdown to occur.

The voltage as a function of time obtained from electrothermal simulations of n-channel MOS transistors under constant current pulsed conditions is shown in Figure 6.7. The gate, substrate and source are at zero volts. At V_{t1} the parasitic npn begins to turn on. The npn is fully on at V_{sp} which is the snapback holding voltage. The temperature begins to rise, which causes the avalanche multiplication to reduce, thereby requiring a higher drain voltage to sustain the npn in the on-state. At V_{t2} the temperature is $\sim$1300 K and the thermally generated holes can now provide the base current, reducing the need for avalanche-generated carriers, and the voltage begins to decrease rapidly. The temperature is a function of the thermal boundary conditions, and this simulation used $R_{th} = 10^6$ K/W in order to reduce internal temperature gradients and speed up the computation. Actual values of R_{th} are closer to 10^4 K/W [Amerasekera93B][Diaz93A]. The 2-D temperature distribution profile for an n-channel MOS transistor at the second breakdown point is shown in Figure 6.8. The temperature is observed to be peaked at the sidewall junction at the drain end, and decreases towards the contacts at the drain and source. Second breakdown is observed to take place when the thermally generated current, I_{th}, is large enough to augment the avalanche-generated current, I_{av}, in supporting the lateral npn in the on-state. As the temperature increases, I_{av} decreases due to the decrease in the impact ionization coefficient, while I_{th} increases. When I_{th} becomes a sufficiently

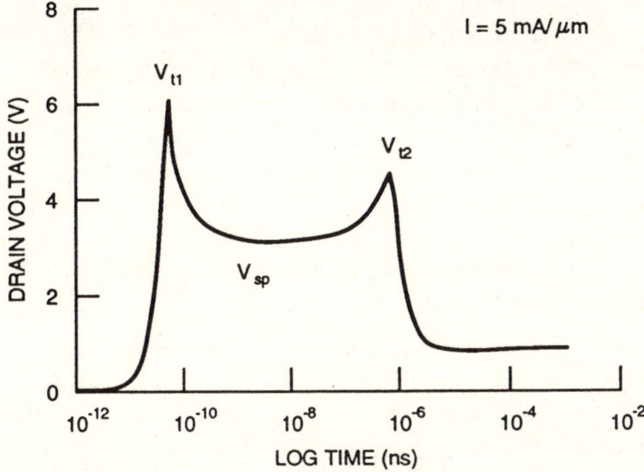

Figure 6.7 Electrothermal simulation of voltage as a function of time for an nMOS transistor with a constant current of 5 mA/μm injected at the drain. The gate, source and substrate are at 0 V.

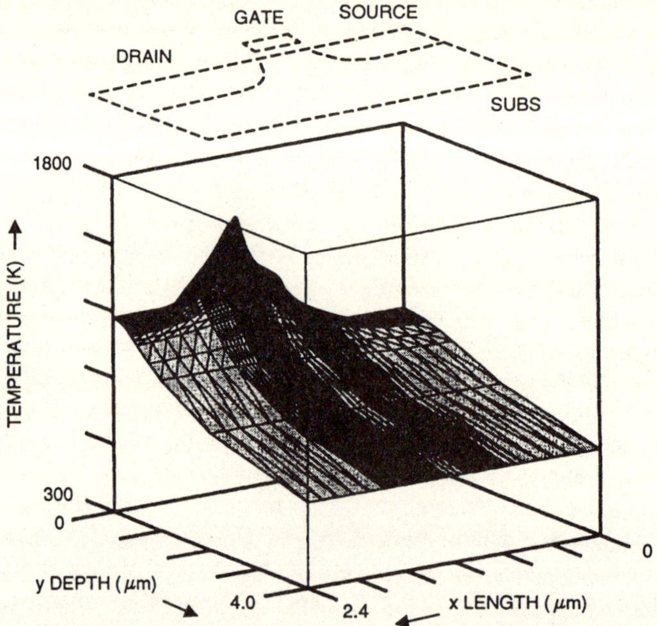

Figure 6.8 2-D temperature distribution in an nMOS transistor at second breakdown obtained from electrothermal device simulation. The injected current is 5 mA/ μm and $R_{th} = 10^6$ K/W at the drain, source, gate and substrate contacts (after [Amerasekera93A]).

large fraction of the base current, then I_{av} is no longer required to increase to support the transistor action and the voltage eventually begins to decrease. The resultant negative resistance region results in instabilities and the onset of thermal second breakdown [Amerasekera94B].

It has also been shown [Yang93] that it is possible to simulate current filamentation under thermal conditions using an electrothermal simulator. Thus, electrothermal simulations appear to be able to duplicate the correct trends required for the simulation of ESD phenomena in semiconductor devices.

While there is a large amount of work that has been done in the area of electrothermal simulations, there is still a need to obtain good correlations between the different phenomena observed experimentally. In developing predictive techniques it is necessary to ensure that correlations with experimental results show the correct trends for both the failure thresholds as well as the times to failure. Not doing so can lead to incorrect results due to inaccuracies which may arise due to grid/mesh effects or problems with convergence [Amerasekera93A].

6.6 CIRCUIT SIMULATIONS

The primary purpose of circuit simulations is to determine whether an ESD protection circuit will operate within the required boundary conditions in terms of current and voltage necessary to prevent damage to the circuit being protected. Once the boundary conditions for failure are known, circuit simulations can provide details about whether these conditions are exceeded for a given stress ·level. Circuit level simulations have been used to develop and optimize novel protection circuits without going through a number of design cycles [Chatterjee91][Duvvury92A]. Such simulation approaches use SPICE and essentially depend on the existing or extracted SPICE parameters for a given protection circuit [Diaz93B][Diaz93C].

More detailed simulations including electrothermal effects are also possible at the circuit level using lumped element models [Beltman90][Kurimoto94]. A three-dimensional thermal scheme is shown in Figure 6.9 [Beltman90]. The thermal resistances and capacitances are calculated from the geometry and the thermal properties of the silicon. The power dissipation is represented by current sources from the electrical scheme. The aluminum contacts can also be represented in the same scheme. In addition, the electrical resistances are taken to be temperature dependent thereby including the intrinsic thermal effects on the device.

Since MOS devices during an ESD event operate as lateral bipolar transistors, equivalent circuit models for these devices need to include the temperature dependence of the bipolar action [Roman70][Scott86][Diaz93C]. Figure 6.10 shows an equivalent circuit for a parasitic bipolar [Scott86]. The avalanche generation of holes which provide the base current is modeled as a current source I_M given by

$$I_M = (M - 1)I_c \qquad (6.48)$$

where M is the multiplication factor and I_c is the collector current. The dependence of M on the applied voltage has been taken as the *Miller* approximation [Miller57][Dutton75][Diaz93C]

$$M = \frac{1}{1 - (V/BV)^n} \qquad (6.49)$$

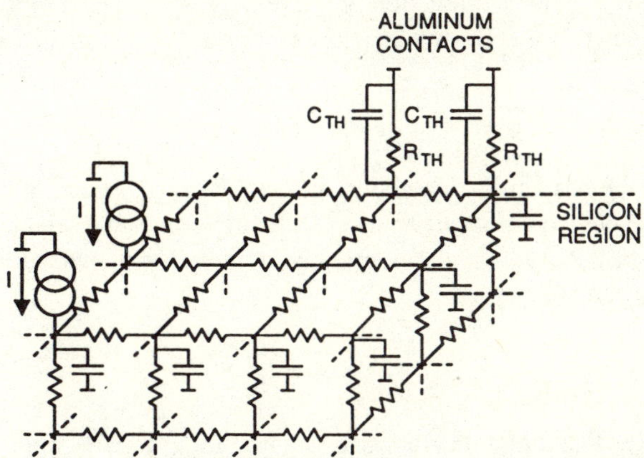

Figure 6.9   3-D thermal network for circuit simulation. Each node corresponds to a dissipating element in the electrical scheme (after [Beltman90]).

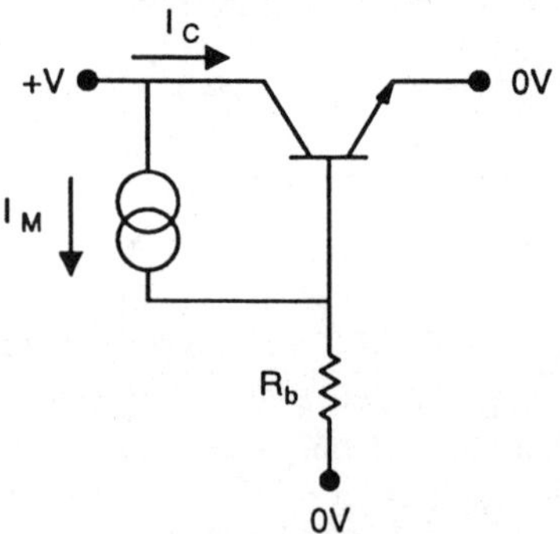

Figure 6.10 Equivalent circuit of a parasitic *npn* transistor with avalanche generation modeled by a current source, I_M (after [Scott86]).

BV is the collector/base junction breakdown voltage and *n* is a power dependence which is $\sim$2 for silicon n^+/p junctions [Miller57]. In reality, as *V* approaches *BV*, *n* approaches 1 and will differ for different junctions. Use of this approximation without exact knowledge of *n* will not provide accurate results, but should still enable a qualitative circuit analysis to be made.

At room temperature, $I_c = M\alpha I_e$ [Reisch92][Luchies94], assuming that recombination in the *p*-base region is negligible and that the thermal current, I_{co} is small. α in this case is the common-base gain which typically approaches 1 in these devices. However, as the temperature increases I_{co} begins to contribute to the base current and I_c becomes

$$I_c = M\alpha I_e + MI_{co} \tag{6.50}$$

Substituting for I_e [Sze81][Muller86] we get

$$I_c(T) = MI_s(T) \cdot \exp\left(\frac{qV_{BE}}{kT}\right) + MI_{co}(T) \tag{6.51}$$

where *T* is the absolute temperature, V_{BE} is the base-emitter voltage, *q* is the electron charge, *k* is the Boltzmann constant and $I_s(T)$ is given by

$$I_s(T) = \frac{qA_e n_i^2(T)D(T)}{N_B} \tag{6.52}$$

A is the base-emitter junction area, $n_i(T)$ is the intrinsic carrier concentration, $D(T)$ is the *effective* minority carrier diffusion constant in the base and N_B is the Gummel number. N_B is the total number of impurities per unit area in the base. I_{co} is approximated by

$$I_{co}(T) = qA_c \frac{D(T)}{\tau} \frac{n_i^2(T)}{N_B} + \frac{qA_c n_i(T)W}{\tau} \tag{6.53}$$

where A_c is the collector-base junction area. Clearly, the temperature dependence of $n_i(T)$ will have the biggest influence on the high temperature value of I_{co} and of $I_s(T)$.

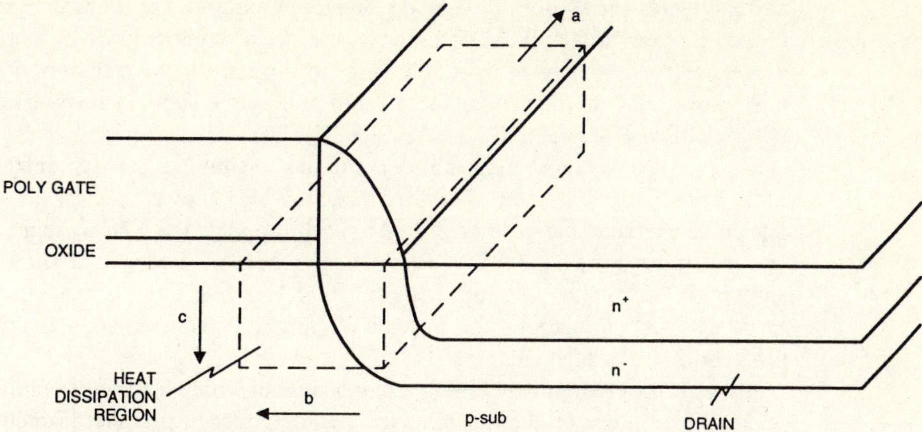

Figure 6.11 Approximation of the heat source in an nMOS transistor (dashed rectangular box).

$n_i(T)$ can be defined by

$$n_i(T) = \sqrt{N_C(T) \cdot N_V(T)} \cdot \exp\left(\frac{-E_g(T)}{2kT}\right) \tag{6.54}$$

N_C and N_V are the effective density of states in the valence and conduction bands, and $E_g(T)$ is the temperature dependent bandgap. $N_C(T)$ and $N_V(T)$ are proportional to $T^{3/2}$. The temperature dependence of n_i is important in determining how $I_c(T)$ varies as the device heats up.

The effect of heating has been included by using approximate solutions to the heat equation to determine the temperature for a given input power and a known heat source [Scott86][Diaz93C]. In MOS structures, the heat source can be assumed to be given by the device width, the depletion width and the junction depth as shown by a, b and c in Figure 6.11 [Amerasekera91]. The variation in gain as a function of temperature is mainly due to reduction of the bandgap at high temperatures. The temperature dependence of the gain is given by [Sze81][Scott86]

$$\beta = \beta_0 \cdot \frac{\exp(-\frac{\Delta E_g}{kT})}{1 + \frac{I_E}{I_{ref}}} \tag{6.55}$$

where ΔE_g is the reduction in the bandgap and I_{ref} is the emitter current at which the current gain begins to decrease. When the device is in the snapback mode, β must satisfy the condition [Scott86]

$$\beta \cdot (M - 1)\eta \geq 1 \tag{6.56}$$

where η is the fraction of the avalanche current that flows across the base-emitter junction. In submicron nMOS devices $\eta \approx 1$.

Second breakdown occurs in the *npn* transistor when the thermally generated component of the base current becomes a larger fraction of the base current, thereby enabling the avalanche-generated component to reduce. As the avalanche-generated current decreases, so does the voltage and a thermally induced negative resistance is initiated [Amerasekera94B]. The onset of second breakdown is not, therefore, associated with the generation of electrons to support the injected current but instead it is associated with the generation of holes required to support the base current. This is a much lower current.

It is possible to combine the above equations to simulate the high current behavior of a lateral *npn* transistor [Diaz93C][Luchies94]. However, there are still many inaccuracies due to the fitting required to accommodate the voltage dependence of M and the temperature dependence of β. The results published to-date show that circuit simulation techniques are improving but more work is required before circuit level electrothermal simulations can provide quantitative and predictive analysis of ESD protection circuit elements.

A major problem in developing a circuit model which comprehends the effects of an ESD-type pulse on the full IC is the large number of parasitics which need to be extracted and computed [Yang93]. For this reason it is not considered feasible to develop a detailed circuit simulator for ESD in a chip. It is possible to get around this issue to some extent by using mixed mode (combined circuit and device) simulators and lumped element thermal models. A reasonably accurate simulator can be pieced together with such methods which will enable good ESD protection circuits to be developed purely from simulations. The work of Chatterjee *et al* showed how SPICE could be used in conjunction with the TMA PISCES device simulator to develop a successful first pass ESD protection circuit, in this case for a BiCMOS IC. The methodology requires that the primary protection circuit element for the technology is first identified. In the example given by Chatterjee, a bipolar *npn* transistor was selected. The device simulator then determines the bias conditions required to obtain a trigger voltage which is less than the maximum clamping voltage required to protect the internal circuitry. Figure 6.12 shows the

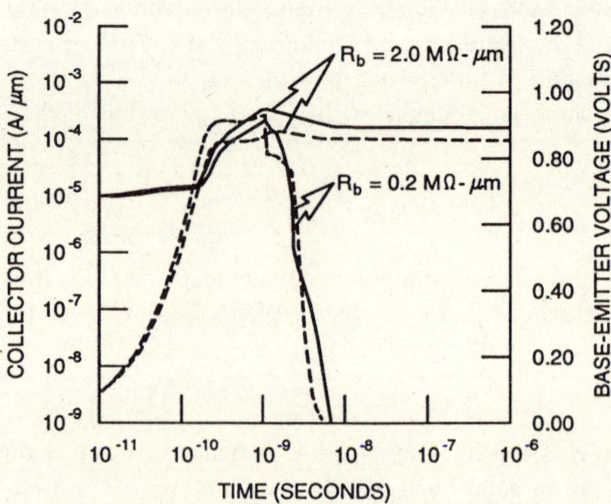

Figure 6.12 Collector current (solid line) and base-emitter voltage (dashed line) for base resistances, R_b, of 0.2 MΩ-μm and 2 MΩ-μm (after [Chatterjee91]). These results are obtained from device simulations.

collector current and the base-emitter voltage of the *npn* transistor as a function of time for different base resistances in response to a 10 V pulse with a rise time of 1 ns. The device simulator is then used to determine the response of the device to pulses with different rise times. ESD pulse rise times vary according to the type of ESD test model used, and the ESD protection circuit must be able to operate within the full range of the tester specifications. The device simulator can also be used for determining the worst-case operating conditions in terms of the bias conditions and rise times. These results are then used as the boundary conditions in determining the operation of the full output/input protection circuit in the SPICE circuit simulator. The response of the circuit to a typical ESD type pulse is obtained for different rise times and peak voltages expected to be seen at the pad. The results of the base-emitter voltage as a function of time for the *npn* as obtained from the circuit simulation are shown in Figure 6.13. Since V_{be} needs to be $\approx$0.6 V for the circuit to operate, the SPICE simulations will enable the designer to identify the limits of operation of his protection and how it responds to variations in circuit or device parameters. A similar methodology was used by Duvvury, Diaz and Haddock [Duvvury92A][Duvvury92B] in developing and optimizing the gate-coupled NMOS transistor as an ESD protection circuit.

Simple circuit simulation methodologies as described here are easy to implement and to follow and require no complex simulation tools. It is, therefore, surprising that they are not used more frequently in ESD protection optimization. One problem is the difficulty in simulating parasitic devices such as the lateral bipolar transistor and the SCR typically used in ESD protection circuits. However, this problem has been addressed by Duvvury and Diaz [Duvvury92A], where PISCES was used to obtain the SPICE parameters of the parasitic elements which were then implemented in the circuit simulator. The simplicity and ease of usage of these circuit simulation techniques should make them attractive to circuit designers. The approach is certainly recommended by the authors at least until more advanced and specialized tools are available for the full-circuit simulation of ESD events on a chip.

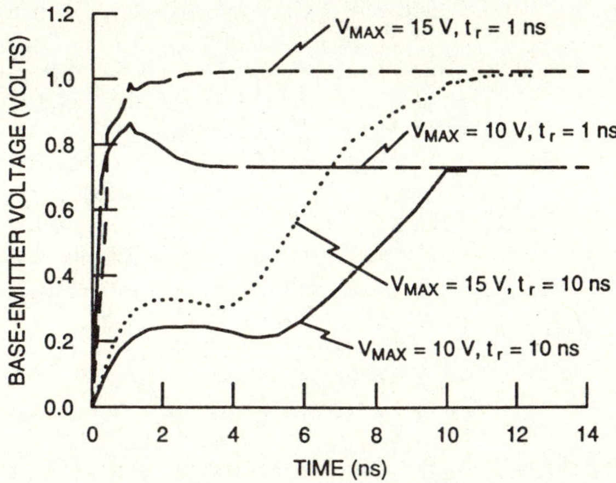

Figure 6.13 Base-emitter voltage response for the *npn* for different rise times t_r and peak voltages V_{MAX}.

REFERENCES

[Abderhalden91]   J. Abderhalden, *Untersuchungen zur Optimierung von Schutzstrukturen gegen elektrostatische Entladungen in integrierten CMOS-Schaltungen*, PhD Thesis, Eidgenossische Technische Hochschule (ETH), Zurich, 1991.

[Adler78]   M. Adler, 'Accurate Calculations of the Forward Drop and Power Dissipation in Thyristors', *IEEE Trans. Elec. Dev.*, ED-25, p. 16–22, 1978.

[Alexander78]   D. R. Alexander, 'Electrical Overstress Failure Modeling for Bipolar Semiconductor Components', *IEEE Trans. Comp. Hyb. Man. Tech.*, CHMT-1, p. 345–353, 1978.

[Alwin77]   V. S. Alwin, D. Navon, L. Turgeon, 'Time-Dependent Carrier Flow in a Transistor Structure under Nonisothermal Conditions', *IEEE Trans. Elec. Dev.*, ED-24, p. 1297–1303, 1977.

[Amerasekera90]   A. Amerasekera, L. J. van Roozendaal, J. Abderhalden, J. J. P. Bruines, L. Sevat, 'An Analysis of Low Voltage ESD Damage in Advanced CMOS Processes', in *Proc. 12th EOS/ESD Symposium*, p. 143–150, 1990.

[Amerasekera91]   A. Amerasekera, L. van Roozendaal, J. Bruines, F. Kuper, 'Characterization and Modeling of Second Breakdown in NMOST's for the Extraction of ESD-Related Process and Design Parameters', *IEEE Trans. Elec. Dev.*, ED-38, p. 2161–2168, 1991.

[Amerasekera92]   A. Amerasekera, W. van den Abeelen, L. J. van Roozendaal, M. Hannemann, P. J. Schofield, 'ESD Failure Modes: Characteristics, Mechanisms and Process Influences', *IEEE Trans. Elec. Dev.*, ED-39, p. 2, 1992.

[Amerasekera93A]   A. Amerasekera, A. Chatterjee, M.-C. Chang, 'Prediction of ESD Robustness of a Process Using 2-D Device Simulations', in *Proc. 31st Int. Rel. Phys. Symp.*, p. 161–167, 1993.

[Amerasekera93B]   A. Amerasekera, M.-C. Chang, J. Seitchik, A. Chatterjee, K. Mayaram, J.-H. Chern, 'Self-Heating Effects in Basic Semiconductor Structures', *IEEE Trans. Elec. Dev.*, ED-40, p. 1836–1844, 1993.

[Amerasekera94A]   A. Amerasekera, R. Chapman, 'Technology Design for High Current and ESD Robustness in a Deep Submicron Process', *Elec. Dev. Lett.*, 15, p. 383–385, 1994.

[Amerasekera94B]   A. Amerasekera, J. Seitchik, 'Electrothermal Behavior of Deep Submicron nMOS Transistors under High Current Snapback (ESD/EOS) Conditions', in *Tech. Dig. IEDM*, p. 446–449, 1994.

[Arkhipov83]   V. I. Arkhipov, E. R. Astvatsaturyan, V. I. Godovitsyn, A. I. Rudenko, *Int. J. Elec.*, 55, p. 395–401, 1983.

[Bardeen50]   J. Bardeen, W. Shockley, 'Deformation Potentials and Mobilities in Nonpolar Crystals', *Phys. Rev.*, 80, p. 72–76, 1950.

[Beltman90]   R. A. M. Beltman, H. van der Vlist, A. J. Mouthaan, 'Simulation of Thermal Runaway during ESD Events', in *Proc. 12th EOS/ESD Symposium*, p. 157–161, 1990.

[Brown72]   W. D. Brown, 'Semiconductor Device Degradation by High Amplitude Current Pulses', *IEEE Trans. Nucl. Sci.*, NS-19, p. 68–75, Dec. 1972.

[Burgess60]   R. Burgess, 'Negative Resistance in Semiconductor Devices', *Canadian Journal of Physics*, 38, p. 369–375, 1960.

[Callen60]   H. B. Callen, *Thermodynamics*, Wiley: New York, 1960.

[Carslaw59]   H. S. Carslaw, J. C. Jaeger, *Conduction of Heat in Solids*, 2nd edn, London: Oxford University Press, 1959.

[Caruso74]   A. Caruso, P. Spirito, G. Vitale, 'Negative Resistance Induced by Avalanche Injection in Bulk Semiconductors', *IEEE Trans. Elec. Dev.*, ED-21, p. 578–586, 1974.

[Caughey67]   D. M. Caughey, R. E. Thomas, 'Carrier Mobilities in Silicon Empirically Related to Doping and Field', *Proc. IEEE*, 55, p. 2192–2193, 1967.

[Chatterjee91]   A. Chatterjee, T. Polgreen, A. Amerasekera, 'Design and Simulation of a 4 kV ESD Protection Circuit for 0.8 μm BiCMOS Process', in *Tech. Dig. IEDM*, p. 913–916, 1991.

[Chen71]   H. C. Chen, W. M. Portnoy, D. K. Ferry, 'Doping Dependence of Second Breakdown in a *p–n* Junction', *Sol. St. Elec.*, 14, p. 747–751, 1971.

[Chryssafis79]   A. Chryssafis, W. Love, 'A Computer-Aided Analysis of One-Dimensional Thermal Transients in *npn* Power Transistors', *Sol. St. Elec.*, 22, p. 249–256, 1979.

[Chynoweth60]   A. Chynoweth, 'Uniform Silicon *p–n* Junctions. II. Ionization Rates for Electrons', *J. Appl. Phys.*, 31, p. 1161–1165, 1960.

[Crowell66]   C. R. Crowell, S. M. Sze, 'Temperature Dependence of Avalanche Multiplication in Semiconductors', *App. Phys. Lett.*, 9, p. 242–243, 1966.

[Daniel90]   S. Daniel, G. Krieger, 'Process and Design Optimization for Advanced CMOS I/O ESD Protection Devices', in *Proc. 12th EOS/ESD Symposium*, p. 206–213, 1990.

[Diaz92]   C. Diaz, S. Kang, C. Duvvury, L. Wagner, 'Electrical Overstress (EOS) Power Profiles: A Guideline to Qualify EOS Hardness of Semiconductor Devices', in *Proc. 14th EOS/ESD Symposium*, p. 88–94, 1992.

[Diaz93A]   C. Diaz, C. Duvvury, S.-M. Kang, 'Studies of EOS Susceptibility in 0.6 μm nMOS ESD I/O Protection Structures', in *Proc. 15th EOS/ESD Symposium*, p. 83–91, 1993.

[Diaz93B]   C. Diaz, C. Duvvury, S.-M. Kang, 'Electrothermal Simulation of Electrical Overstress in Advanced nMOS ESD I/O Protection Circuits', in *Tech. Dig. IEDM*, p. 899–902, 1993.

[Diaz93C]   C. Diaz, 'Modeling and Simulation of Electrical Overstress Failures in Input/Output Protection Devices of Integrated Circuits', PhD Thesis, University of Illinois at Urbana-Champaign, 1993.

[Dutton75]   R. W. Dutton, 'Bipolar Transistor Modeling of Avalanche Generation for Computer Circuit Simulation', *IEEE Trans. Elec. Dev.*, ED-22, p. 334–338, 1975.

[Duvvury88]   C. Duvvury, R. Rountree, O. Adams, 'Internal Chip ESD Phenomena beyond the Protection Circuit', in *Proc. 26th IRPS*, p. 19–25, 1988.

[Duvvury92A]   C. Duvvury, C. Diaz, 'Dynamic Gate-Coupled NMOS for Efficient Output ESD Protection', in *Proc. 30th IRPS*, p. 141–150, 1992.

[Duvvury92B]   C. Duvvury, C. Diaz, T. Haddock, 'Achieving Uniform nMOS Power Distribution for Improving ESD/EOS Reliability', in *Tech. Dig. IEDM*, p. 131–134, 1992.

[Dwyer90]   V. M. Dwyer, A. J. Franklin, D. S. Campbell, 'Thermal Failure in Semiconductor Devices', *Sol. St. Elec.*, 33, p. 553–560, 1990.

[Gaur76]   S. Gaur, D. Navon, R. Teerlinck, 'Transistor Design and Thermal Stability', *IEEE Trans. Elec. Dev.*, ED-20, p. 527–534, 1976.

[Geballe55]   T. H. Geballe, G. W. Hull, 'Seebeck Effect in Silicon', *Phys. Rev.*, 98, p. 940–947, 1955.

[Ghandhi77]   S. K. Ghandhi, *Semiconductor Power Devices*, New York: Wiley, 1977.

[Grant73]   W. N. Grant, 'Electron and Hole Ionization Rates in Epitaxial Silicon at High Electric Fields', *Sol. St. Elec.*, 16, p. 1189–1203, 1973.

[Grutchfield66]   H. B. Grutchfield, T. J. Moutoux, 'Current Mode Second Breakdown in Epitaxial Planar Transistors', *IEEE Trans. Elec. Dev.*, ED-13, p. 743, 1966.

[Hower70]   P. L. Hower, V. G. K. Reddi, 'Avalanche Injection and Second Breakdown in Transistors', *IEEE Trans. Elec. Dev.*, ED-17, p. 320–335, 1970.

[Jacoboni77]   C. Jacoboni, C. Canali, G. Ottaviani, A. Quaranta, 'A Review of Some Charge Transport Properties of Silicon', *Sol. St. Elec.*, 20, p. 77–89, 1977.

[Khurana66]   B. S. Khurana, T. Sugano, H. Yanai, 'Thermal Breakdown in Silicon *p–n* Junction Devices', *IEEE Trans. Elec. Dev.*, ED-13, p. 763–770, 1966.

[Koyanagi77]   K. Koyanagi, K. Hane, T. Suzuki, 'Boundary Conditions between Current Mode and Thermal Mode Second Breakdown in Epitaxial Planar Transistors', *IEEE Trans. Elec. Dev.*, ED-24, p. 672–678, 1977.

[Krabbenborg91]   B. Krabbenborg, R. Beltman, P. Wolbert, T. Mouthan, 'Physics of Electro-Thermal Effects in ESD Protection Devices', in *Proc. 13th EOS/ESD Symposium*, p. 98–103, 1991.

[Krakauer92]   D. Krakauer, K. Mistry, 'ESD Protection in a 3.3 V Sub-Micron Silicided CMOS Technology', in *Proc. 14th EOS/ESD Symposium*, p. 250–257, 1992.

[Krieger87]   G. Krieger, 'Thermal Response of Integrated Circuit Input Devices to an Electrostatic Energy Pulse', *IEEE Trans. Elec. Dev.*, ED-34, p. 877–882, 1987.

[Kurimoto94]   K. Kurimoto, K. Yamashita, I. Miyanaga, A. Hori, S. Odanaka, 'An Electrothermal Circuit Simulation Using an Equivalent Thermal Network for Electrostatic Discharge', in *Proc. VLSI Symp. on Tech.*, p. 127–128, 1994.

[Lin90]   D. L. Lin, 'Thermal Breakdown of VLSI by ESD Pulses', in *Proc. 29th IRPS*, p. 281–287, 1990.

[Luchies94]   J. Luchies, C. de Kort, J. Verweij, 'Fast Turn-on of an NMOS ESD Protection Transistor; Measurements and Simulations', in *Proc. 16th EOS/ESD Symposium*, p. 266–272, 1994.

[Mathews80]   D. Mathews, 'Some Design Criteria for Avoiding Second Breakdown in Bipolar Devices', in *Proc. 2nd EOS/ESD Symposium*, p. 117–121, 1980.

[Mayaram91]   K. Mayaram, J.-H. Chern, L. Arledge, P. Yang, 'Electrothermal Simulation Tools for Analysis and Design of ESD Protection Devices', in *Tech. Dig. IEDM*, p. 909–912, 1991.

[MEDICI92]   *MEDICI: 2-Dimensional Device Simulation Program*, Technology Modeling Associates Inc., Palo Alto, California, 1992.

[Melchior64]   H. Melchior, M. J. O. Strutt, 'Secondary Breakdown in Transistors', *Proc. IEEE (Correspondence)*, 52, p. 439–440, 1964.

[Miller57]   S. L. Miller, 'Ionization Rates for Holes and Electrons in Silicon', *Phys. Rev.*, 105, p. 1246–1249, 1957.

[Muller86]   R. S. Muller, T. I. Kamins, *Device Electronics for Integrated Circuits*, 2nd edn, New York: Wiley, 1986.

[Nienhuis66]   R. J. Nienhuis, 'Second Breakdown in the Forward and Reverse Base Current Region', *IEEE Trans. Elec. Dev.*, ED-13, p. 655–662, 1966.

[Okuto75]   Y. Okuto, C. R. Crowell, 'Threshold Energy Effect on Avalanche Breakdown Voltage in Semiconductor Junctions', *Sol. St. Elec.*, 18, p. 161–168, 1975.

[Orvis83]   W. J. Orvis, C. F. McConaghy, J. H. Yee, G. H. Khanaka, L. C. Martin, D. H. Lair, 'Modeling and Testing for Second Breakdown Phenomena', in *Proc. 5th EOS/ESD Symposium*, p. 108–117, 1983.

[Overstraeten70]   R. van Overstraeten, H. de Man, 'Measurement of the Ionization Rates in Diffused Silicon *p–n* Junctions', *Sol. St. Elec.*, 13, p. 585–608, 1970.

[Pierce88]   D. G. Pierce, W. Shiley, B. D. Mulcahy, K. E. Wagner, M. Wunder, 'Electrical Overstress Testing of a 256K UVEPROM to Rectangular and Double Exponential Pulses', in *Proc. 10th EOS/ESD Symposium*, p. 137–146, 1988.

[Polgreen89]   T. Polgreen, A. Chatterjee, 'Improving the ESD Failure Threshold of Silicided nMOS Output Transistors by Ensuring Uniform Current Flow', in *Proc. 11th EOS/ESD Symposium*, p. 167–174, 1989.

[Popescu70]   C. Popescu, 'The Thermal Runaway Mechanism of Second Breakdown Phenomenon', *Sol. St. Elec.*, 13, p. 887–901, 1970.

[Reisch92]   M. Reisch, 'On Bistable Behavior and Open-Base Breakdown of Bipolar Transistors in the Avalanche Regime – Modeling and Applications', *IEEE Trans Elec. Dev.*, ED-39, p. 1398–1409, 1992.

[Ridley63]   B. K. Ridley, 'Specific Negative Resistance in Solids', *Proc. Phys. Soc.*, 82, p. 954–966, 1963.

[Roman70]   G. Roman, 'A Model for Computation of Second Breakdown in Transistors', *Sol. St. Elec.*, 13, p. 961–980, 1970.

[Runyan65]   W. B. Runyan, *Silicon Semiconductor Technology*, New York: McGraw-Hill, 1965.

[Scarlett63]   R. M. Scarlett, W. Shockley, R. H. Haitz, 'Thermal Instabilities and Hot Spots in Junction Transistors', in *Physics of Failure in Electronics*, ed. M. F. Goldberg and J. Vaccaro, Baltimore, MD: Spartan Books, 1963, p. 194–203.

[Schafft62]   H. A. Schafft, J. C. French, 'Second Breakdown in Transistors', *IEEE Trans. Elec. Dev.*, ED-9, p. 129–136, 1962.

[Schafft66]   H. A. Schafft, J. C. French, 'A Survey of Second Breakdown', *IEEE Trans. Elec. Dev.*, ED-13, p. 613–618, 1966.

[Scott86]   D. Scott, G. Giles, J. Hall, 'A Lumped Element Model for Simulation of ESD Failures in Silicided Devices', in *Proc. 8th EOS/ESD Symposium*, p. 41–47, 1986.

[Selberherr84]   S. Selberherr, *Analysis and Simulation of Semiconductor Devices*, New York: Springer-Verlag, 1984.

[Shaw92]   M. P. Shaw, V. V. Mitin, E. Schöll, H. L. Grubin, *The Physics of Instabilities in Solid State Electron Devices*, New York: Plenum, 1992.

[Smith73]   W. B. Smith, D. H. Pontius, P. P. Budenstein, 'Second Breakdown and Damage in Junction Devices', *IEEE Trans. Elec. Dev.*, ED-20, p. 731–744, 1973.

[Speakman74]   T. S. Speakman, 'A Model for the Failure of Bipolar Silicon Integrated Circuits Subjected to Electrostatic Discharge', in *Proc. 12th IRPS*, p. 60–69, 1974.

[Steele62]   M. C. Steele, K. Ando, M. A. Lampert, 'Avalanche Breakdown Double Injection Induced Negative Resistance of Semiconductors', *J. Phys. Soc. Jap.*, 17, p. 1729–1736, 1962.

[Stratton72]   R. Stratton, 'Semiconductor Current-Flow Equations (Diffusion and Degeneracy)', *IEEE Trans. Elec. Dev.*, ED-19, p. 1288–1300, 1972.

[Sze81]   S. M. Sze, *Physics of Semiconductor Devices*, 2nd edn, New York: Wiley, 1981.

[Tasca70]   D. M. Tasca, 'Pulse Power Failure Modes in Semiconductors', *IEEE Trans. Nucl. Sci.*, NS-17, p. 364–372, Dec. 1970.

[Tasca72]   D. M. Tasca, J. C. Peden, J. Miletta, 'Non-Destructive Screening for Thermal Second Breakdown', *IEEE Trans. Nucl. Sci.*, NS-19, p. 57–67, 1972.

[Tauc57]   J. Tauc, A. Abraham, 'Thermal Breakdown in Silicon p–n Junctions', *Phys. Rev.*, 108, p. 936–937, 1957.

[Wachutka90]   G. Wachutka, 'Rigorous Thermodynamic Treatment of Heat Generation and Conduction in Semiconductor Device Modeling', *IEEE Trans. Comp. Aid. Des.*, CAD-9, p. 1141–1149, 1990.

[Ward76]   A. L. Ward, 'An Electrothermal Model of Second Breakdown', *IEEE Trans. Nucl. Sci.*, NS-23, p. 1679–1684, 1976.

[Ward77]   A. L. Ward, 'Studies of Second Breakdown in Silicon Diodes', *IEEE Trans. Parts, Hybrids and Packaging*, PHP-13, p. 361–368, Dec. 1977.

[Wunsch68]   D. C. Wunsch, R. R. Bell, 'Determination of Threshold Failure Levels of Semiconductor Diodes and Transistors Due to Pulse Voltages', *IEEE Trans. Nucl. Sci.*, NS-15, p. 244, Dec. 1968.

[Yang93]   P. Yang, J.-H. Chern, 'Design for Reliability: The Major Challenge for VLSI', *Proc. IEEE*, 81, p. 730–744, 1993.

[Yee82]   J. Yee, W. Orvis, L. Martin, J. Peterson, 'Modeling of Current and Thermal Mode Second Breakdown Phenomena', in *Proc. 4th EOS/ ESD Symposium*, p. 76–81, 1982.

7

INFLUENCE OF PROCESSING ON ESD

7.1 INTRODUCTION

The fabrication process has a strong impact on the ESD sensitivity of protection circuits. A circuit which functions extremely well in one technology may show a very poor performance in another technology. This is especially true when translating ESD protection circuits through technology shrinks. Hence, an understanding of the key process parameters influencing ESD is essential to the development of ESD protection circuit design methodology. While significant improvements have been made in this understanding, there are still a number of fundamental issues which need to be addressed with regard to the influence of the technology on ESD performance.

There are many reasons for the strong process impact on ESD. Most ESD protection circuits depend on the action of various parasitic elements to provide the necessary current shunting and voltage clamping. Even in circuits with current paths defined by non-parasitic elements, such as *pn* diodes, parasitic elements in the internal chip circuitry will eventually trigger and influence the ESD behavior. Such circuits usually have reverse-biased junctions where heat dissipation can lead to eventual thermal breakdown. In the previous chapter, Figure 6.1(A) showed a cross-section of an nMOS transistor operating as a parasitic lateral *npn* device. The collector junction is reverse-biased and the base current to sustain the *npn* in the on-state is provided by avalanche generation in the reverse-biased collector (drain) junction. The temperature rise in the junction and the onset of thermal breakdown are dependent on the power density, $J \cdot E$ (W/cm^3), in the junction. The boundary condition for damage is the eventual collapse of the current in a localized region leading to the formation of melt filaments. Both the current density, J, and the electric field, E, are functions of the doping profiles of the junctions and the substrate and, therefore, process dependent [Shabde84][Duvvury89][Ohtani90][Amerasekera90].

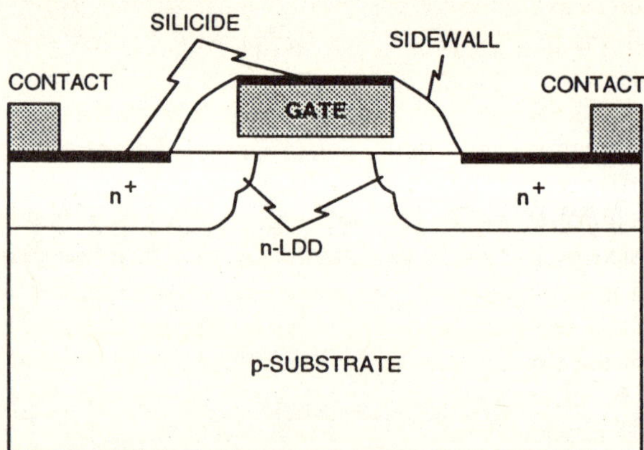

Figure 7.1   Schematic cross-section of an nMOS transistor with LDD and silicide.

The resistance between the contact and the collector junction where the heating takes place will influence the onset of current localization and failure. Hence, the contact resistance and the source/drain sheet resistance are important process parameters. Advanced CMOS technologies introduced silicide-clad source and drain diffusions to reduce the resistance and improve the speed of the transistors [Lau82]. Figure 7.1 shows a cross-section of an nMOS transistor indicating the location of the silicide cladding. The result was a drastic reduction in the ESD thresholds of previously high-performing protection circuit elements [Duvvury85][McPhee86][Chen86].

Resistance plays a major role in defining the voltage clamping limits of a number of ESD protection circuits. Resistors can be directly used as part of protection circuits using secondary and primary circuit elements. Changes in the resistance value can lead to the primary circuit not turning on before the voltage clamp limit is reached thus resulting in damage to the device being protected or the secondary circuit. In *pn* diode protection elements, the resistance of the *pn* diode determines the maximum current (and thus the ESD voltage) that the circuit can handle before the voltage clamp limit is reached [Voldman93][Dabral93][Voldman94].

In this chapter the phenomena and mechanisms involved in each of these effects will be presented. First, experimental results showing the effects of variations in the process parameters on ESD performance are shown, and then the phenomena and mechanisms involved will be discussed. The importance of understanding the process design windows in defining and designing ESD protection circuits will be shown. All ICs are eventually packaged, and the main effects of the package on the ESD performance are discussed at the end of this chapter.

7.2 SOURCE/DRAIN JUNCTION EFFECTS

In technologies with feature sizes greater than 2 μm, abrupt junctions were formed with arsenic implantation. These junctions were of the order of 0.5 μm deep and had high doping concentrations. Typically, avalanche breakdown occurred uniformly through the junction depth, and the parasitic bipolar action utilized the entire junction sidewall. The first indications of the importance of the source/drain junction profiles to ESD performance came in sub-2 μm technologies with the introduction of graded junctions and Lightly Doped Drain (LDD) regions in an attempt to reduce the hot carrier sensitivity [Shabde84][Duvvury86]. In Figure 7.1 the cross-section of the nMOS transistor shows the LDD region. In a typical LDD structure, the LDD region is formed by phosphorus implantation with doses ranging from 10^{13}/cm^2 to 10^{14}/cm^2. It is also possible to use arsenic as the implant species for the LDD. The LDD implant is self-aligned to the polysilicon gate edge. The main source/drain (S/D) implant usually consists of arsenic with doses in the 2×10^{15}/cm^2 to 5×10^{15}/cm^2 range. In order to ensure that the LDD region is not overdoped by the S/D implant, the S/D implant is self-aligned to a spacer deposited against the gate as shown in Figure 7.1. The spacer can be between 0.1 μm and 0.25 μm in thickness depending on the feature size of the technology. Both silicon dioxide and silicon nitride spacers are used in advanced technologies.

Shabde *et al* [Shabde84] in one of the earliest papers evaluating the effect of source/drain junction grading on high current robustness and ESD thresholds, showed that as the phosphorus implant dose is increased from 0 to 4×10^{13}/cm^2, the current required to cause breakdown decreases. The ESD sensitivity of a given protection circuit, measured in terms of the probability of damage at ESD levels of 2000 V, is shown to decrease from 0% for no LDD implant to 44% for a 4×10^{13}/cm^2 implant.

The influence of LDD junctions on ESD performance was also studied by Duvvury *et al* [Duvvury86] for both field oxide type lateral *npn* devices and standard nMOS transistors used in output buffers. Figure 7.2 shows the ESD performance for three process variations. The 2 μm process had abrupt junctions while the 1 μm processes had LDD junctions. In addition, the effect of silicide vs.

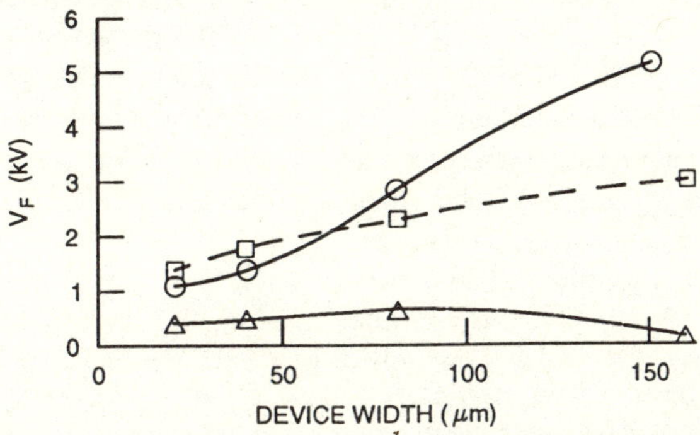

Figure 7.2 ESD failure voltage as a function of device width for thick oxide *n*-type devices from three different source/drain processes (after [Duvvury86]).

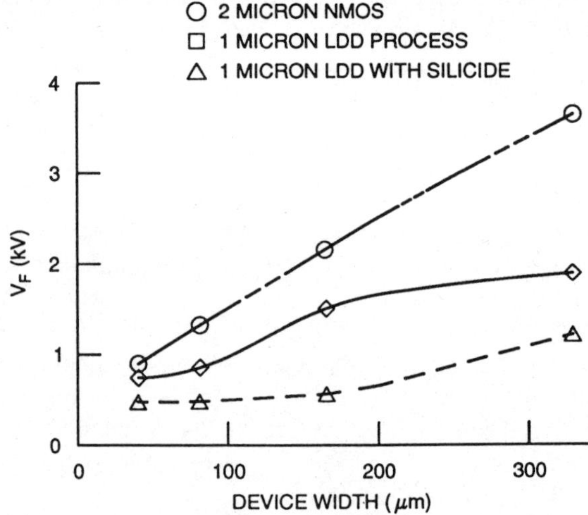

Figure 7.3 ESD failure voltage as a function of device width for thin oxide nMOS transistors from three different source/drain processes (after [Duvvury86]).

non-silicide is shown for the 1 μm process. The ESD capability of field oxide devices in the LDD process was down by a factor of 3 from those in a non-LDD process. The equivalent ESD failure threshold voltage normalized to the width of the device, ESD V/μm, for the LDD device was 15 V/μm, compared to 45 V/μm for the abrupt junction device. The results for nMOS transistors with the same process variations are shown in Figure 7.3. The difference in performance between the LDD and the abrupt junction devices was $\sim$1.5. The LDD devices had a typical capability of 10 V/μm compared to 15 V/μm for the abrupt junction devices. In a later paper, the effects of the arsenic S/D implants on the ESD performance were also shown, indicating that the junction grading had a negative impact on ESD [Duvvury89]. Results showed that a graded junction device with a phosphorus implant dose of 5×10^{14}/cm^2 had an average ESD level of about 10 V/μm, compared to almost 15 V/μm for devices with both phosphorus and a 2×10^{15}/cm^2 arsenic implant. These devices did not have separate LDD implants. Devices with higher phosphorus doping levels of 1.2×10^{15}/cm^2 and no arsenic had higher V/μm levels but were still not as good as devices with the arsenic implant. Results comparing double-diffused drain structures, where the arsenic and phosphorus are both implanted into the S/D area, with LDD structures and graded junction devices are also given by Chen [Chen88]. He shows that double-diffused devices are better by a factor of almost 3 than equivalent LDD devices.

The variation of ESD performance with the phosphorus LDD dose in general seems to indicate that for a non-silicided 0.5 μm process, as the phosphorus dose is increased from 10^{13}/cm^2 to 4×10^{13}/cm^2, the ESD performance increases as shown in Figure 7.4 [Ishizuka94]. However, earlier work [Shabde84] showed that in the same range the ESD performance decreased. Similar trends were observed by Ohtani and Yoshida [Ohtani90] for phosphorus doses from 4×10^{12}/cm^2 to 5×10^{14}/cm^2.

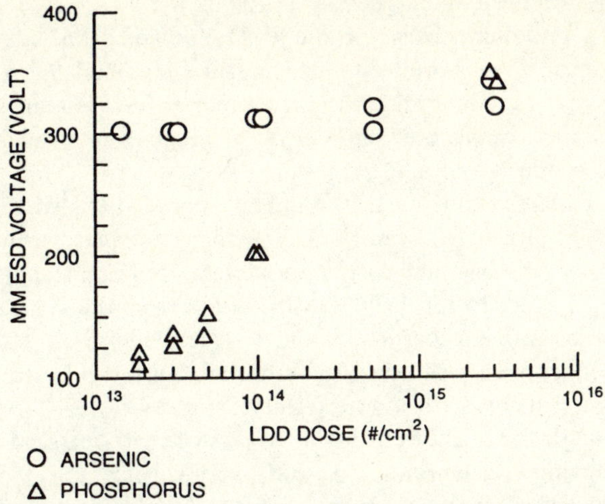

Figure 7.4   MM ESD threshold voltage as a function of As LDD and P LDD dose (after [Ishizuka94]).

The 4×10^{12}/cm^2 dose and the 5×10^{14}/cm^2 dose were shown to give high ESD thresholds while an intermediate 5×10^{13}/cm^2 dose gave a low ESD threshold. It has been shown [Ishizuka94] that devices with a phosphorus LDD dose are much weaker than those using the same arsenic LDD dose. Higher phosphorus LDD doses do indeed improve the ESD performance of these devices.

The difference in the results may be explained by the influence of the p-well doping profile which would not be the same for the different processes. Clearly a higher p-well concentration will result in a lower avalanche breakdown voltage for the same LDD dose, and will certainly reduce the gain of the lateral *npn* transistor. A non-uniform p-well profile will favor current paths which may be different from those of a uniform p-well. Unfortunately, it is not easy to identify the exact p-well profiles in the active transistor regions, especially in submicron technologies [Rafferty93]. But it could be responsible for some of the variations observed in the reported results.

In addition to the work on LDDs, tresults have been reported on the effect of the arsenic S/D implant dose on ESD [Chaine92]. These results need to be compared with the work on graded junctions discussed above [Chen88][Duvvury89]. As the arsenic S/D dose is reduced, the resistance of the drain increases providing more ballast for the drain junction. Thus current filamentation is inhibited and higher second breakdown thresholds will be achieved.

One approach to improving the ESD performance of devices with LDD has been to use the double-diffused drain (DDD) where a high phosphorus dose (5×10^{14}/cm^2 to 10^{15}/cm^2) is implanted into the S/D. The intention is to make the junction deeper as well as to overdope the lightly doped region, thus creating a drain profile similar to those of the abrupt junction technologies [Daniel90][Amerasekera90][Wei92]. Results have shown that the additional (ESD) implant can improve ESD performance by more than a factor of 2, even in a fully-silicided process

[Amerasekera90][Amerasekera91]. As in the LDD case, different authors report different optimum doses for the ESD implant. This could also be attributed to differences in the p-well doping concentrations and the p-well doping profiles. A higher p-well concentration requires a higher ESD implant dose than a lower p-well concentration because of the reduced depletion region width and its impact on the turn-on of the lateral *npn* transistor.

The problem with the ESD implant approach is that the additional phosphorus implant tends to significantly change transistor performance, especially the short-channel effects and hot carrier reliability. Hence, the implementation requires an additional mask to block the ESD implant from the critical transistors on the chip and adds to process complexity and cost. It is best to try to integrate ESD robustness into the technology as a whole by optimizing the S/D and LDD doses for ESD as well as transistor performance [Amerasekera94A].

Although there has been much work in recent years on the effect of the source/drain doping concentrations on ESD performance, the underlying mechanisms are still not very clear. Most analyses have concluded that the improvement with higher doping concentration is due to either deeper junctions or more uniform current flow through the depth of the junction. Another reason may be that the lower doped junctions go into second breakdown at lower temperatures.

We have suggested in this book and elsewhere [Amerasekera94A][Amerasekera94C] that source/drain engineering influences the turn-on behavior of the lateral *npn* transistor. The interaction between the drain depletion region and the source determines whether the conduction mechanism takes place at the surface or below it. The second breakdown trigger current I_{t2} as a function of nMOS transistor drive current I_{drive} for two different LDD processes, one with a feature size (nominal gate length, L) of 0.5 μm and the other with a feature size of 0.35 μm, is shown in Figure 7.5 [Amerasekera94B]. I_{drive} is measured at the nominal gate voltage, V_G, and the

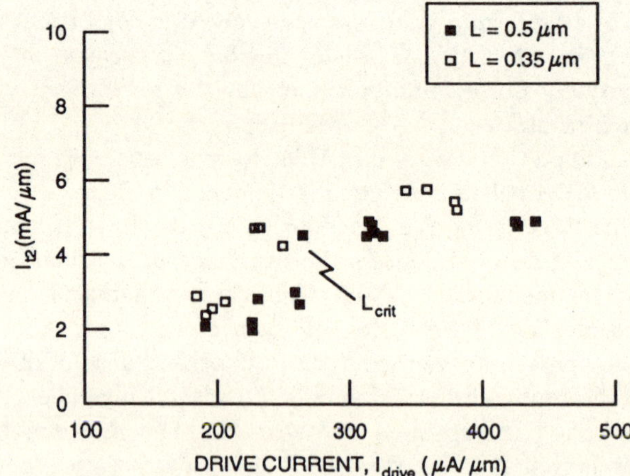

Figure 7.5 I_{t2} as a function of drive current for two different processes with feature sizes of 0.5 μm and 0.35 μm (after [Amerasekera94B]). L_{crit} is the gate length at which I_{t2} begins to decrease.

nominal drain voltage, V_D. For the 0.5 μm process V_G/V_D is 3.3 V/3.3 V, and for the 0.35 μm process, V_G/V_D is 2.5 V/2.5 V. Variations in I_{drive} are obtained by changing the polysilicon gate length, L_G. We see that I_{t2} remains constant until the L equals a critical value L_{crit}. I_{t2} begins to decrease when $L > L_{crit}$ suggesting that the interaction between the drain and source is important to the breakdown mechanism. This interaction is strongly influenced by both the source/drain doping profiles and the well doping profile. In the 0.5 μm process $L_{crit} = 1.2$ μm while in the 0.35 μm process $L_{crit} = 0.65$ μm.

We see that there are still sufficiently conflicting data for each of the different theories discussed here for some doubt to exist as to the exact mechanisms involved in the influence of the doping profiles on ESD. As technologies are scaled to the deep submicron level and beyond (sub-0.25 μm), the need to understand the precise mechanisms involved will be greater. This area will be one of the major focuses of ESD work in the coming years.

7.3 GATE OXIDES

The purpose of an ESD protection circuit is to clamp the voltage at an input gate or output buffer, so that the gate oxide breakdown voltage (BV_{ox}) is not exceeded. Under pulsed conditions oxides can withstand higher electric fields than under steady-state DC conditions [Bridgewood85][Amerasekera86][Fong87]. Although a 100 Å gate oxide may have a $BV_{ox} \approx 10$ V under DC conditions, the pulsed breakdown may be as high as 20 V. In general, the avalanche breakdown voltage of a drain junction, V_{av}, is lower than BV_{ox} as shown in Figure 7.6 for a 0.25 μm process [Amerasekera94B]. However, as the oxide thickness decreases without additional source/drain engineering, it is possible that V_{av} becomes greater than BV_{ox} and gate oxide rupture will occur. This failure mode has been observed in processes with oxide thicknesses of 175 Å [Amerasekera92].

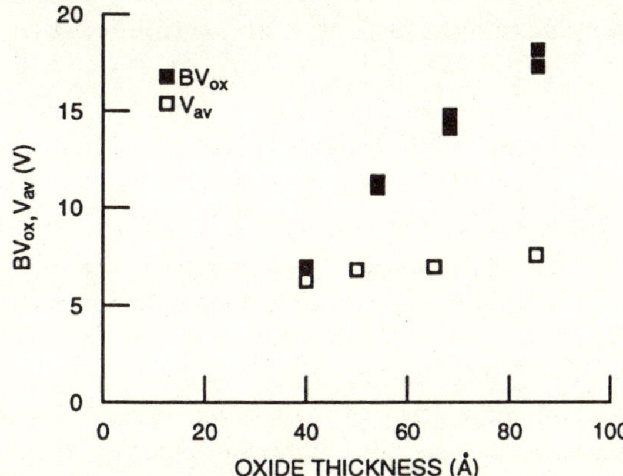

Figure 7.6 Oxide breakdown voltage, BV_{ox}, and junction avalanche breakdown voltage, V_{av}, as a function of oxide thickness for a 0.25 μm CMOS process (after [Amerasekera94B]).

Most clamping devices require that a snapback action of some form eventually takes place, to cause the voltage to fall back to allow suitable margins between V_{av} and BV_{ox} even at current levels of 2 A and greater. However, some processing options can decrease the efficiency of the snapback action. For example, decreasing the epitaxial thickness t_{epi} can raise the trigger voltages and currents of the SCRs used as protection circuits to levels greater than BV_{ox}. Thinner epitaxial layers also make it harder for snapback action to take place in nMOS transistors, and in some cases can eliminate snapback completely. The result is that BV_{ox} is reached at low ESD levels, leading to oxide breakdown.

Another cause of oxide breakdown is when the stress pulse has a very high current level for very short durations, as in the case of the Charged Device Model (CDM) test method. Current levels of the order of 10 A are typical for a 1500 V CDM stress level according to the present industry standard. Under these conditions on-resistance of the protection device in snapback may not be able to prevent the voltage across the oxide from reaching BV_{ox} and oxide breakdown will be observed [Fukuda88]. The reason why oxide breakdown occurs before thermal damage is that the duration of the pulse is less than 1 ns, which does not allow the temperature in the device to reach levels required for thermal breakdown before the voltage reaches BV_{ox}. There is also speculation that the pulse duration is less than the time required for the devices to turn on. However, most parasitic devices in advanced CMOS processes trigger at around 100 ps, and this is not necessarily a primary concern. A suitable clamping device needs to be selected which has a low enough on-resistance that the voltage does not reach BV_{ox} at current levels close to 10 A and durations of ~ 1 ns.

By virtue of its nature, the ESD stress itself is a good screen for defective oxides, and oxides with high defect densities or low charge to breakdown (Q_{bd}) levels will be most sensitive to ESD failure. For good quality oxides as typically found in most processes today, oxide breakdown should not be a problem down to oxide thicknesses of 50 Å. As oxide thicknesses become less than 50 Å, it is important that V_{av} and BV_{ox} are characterized and that a suitable margin is ensured between V_{av} and BV_{ox}. Therefore, ESD robustness needs to be one of the boundary conditions for transistor and process designs in ultra-small devices.

7.4 CONTACTS AND SILICIDATION

In the early 1980s, the contacts were considered to be the weakest link in the protection circuit since the most extensive damage was usually observed at the contacts [DeChiaro81][Maloney86][Duvvury86][Strauss87]. It was first proposed that contact damage was the result of electrothermomigration [DiChiaro81], which was metal migration into the silicon as a result of a combination of the temperature gradient and the electric field between the contact and the hot spot. While this explanation is still used by a number of people to explain these failure modes, Pierce [Pierce85] in a detailed analysis showed that electrothermomigration was not physically possible under ESD conditions, given the time durations of the ESD event. He proposed that metalization burnout and contact damage may actually be a secondary failure mechanism following second breakdown at the junction. The current filamentation which follows second breakdown leads to very high local temperatures and eventual melting of the contact.

The aluminum–silicon eutectic temperature is about 500°C, and when the contact reaches this temperature significant metal migration into the contact will occur. It is possible that when the junction temperature is about 1000°C the thermal gradient between the junction edge and the contact will result in temperatures approaching 500°C at the contact. This explanation of the contact damage mechanism led to design rules specifying the minimum distance between the contact and the junction edge.

In submicron technologies, contacts usually have barrier metals such as titanium-tungsten (TiW) to prevent contact migration under normal operating conditions. TiW has a much higher eutectic temperature ($>1000°C$) which is closer to the melt temperature of silicon, and hence contact spiking observed in older processes now occurs more rarely. In the event that contact spiking is observed in a technology with barrier metals in the contacts, it is probably related to poor step coverage of the TiW layer in the contact [Amerasekera92]. In this case the WAl_{12} alloy formed at the top barrier interface can extend all the way to the silicon and Al–Si interdiffusion is then possible during normal processing [Chang88]. After current localization occurs, the high temperatures can result in the Al forming a spike through the diffusion.

In silicided processes, the source and drain diffusions are clad in titanium silicide, $TiSi_2$, with the intention of reducing the contact resistance [Lau82]. The result was also a dramatic reduction in the ESD performance of protection circuits which functioned well in non-silicided processes, as first shown by Duvvury *et al* [Duvvury86] and confirmed in further studies [Scott86][Wilson87][Chen88]. Figures 7.2 and 7.3 showed the effect of silicides on the ESD performance of thick oxide and thin oxide nMOS devices in a 1 μm process. Silicided devices were between $3\times$ and $6\times$ worse than the non-silicided devices. The increased sensitivity in nMOS output buffers and protection circuits was identified as being caused by the silicide cladding of the source/drain diffusions and not related to the silicide on the polysilicon gates.

From the ESD viewpoint, the primary effect of the silicide cladding on the diffusions is to bring the contact closer to the gate and the diffusion edge. The consequence is that under high current conditions, the ballasting resistance between the contact and the hot spot is reduced. Hence, once a hot spot is initiated at the diffusion edge, there is very little resistance to prevent current localization through the hot spot. When the temperature at the $TiSi_2$ contact reaches 1000°C, the $TiSi_2$ begins to decompose, interact with the silicon, or both, in a similar manner to that of Al at the eutectic temperature. The higher critical temperature indicates that damage to the silicide itself is not the principal cause of failure.

Rountree [Rountree88] showed, Figure 7.7, that the maximum current density in silicided diffusions occurs at the drain and source sidewall edges. This leads to a higher power dissipation at these regions and the damage, therefore, occurs at these edges. It is possible to improve the current distribution in the silicide by changing the contact transfer length, L_c. L_c characterizes the distance over which the current moves from the silicide into the diffusion [Scott87]. Higher values of L_c force the

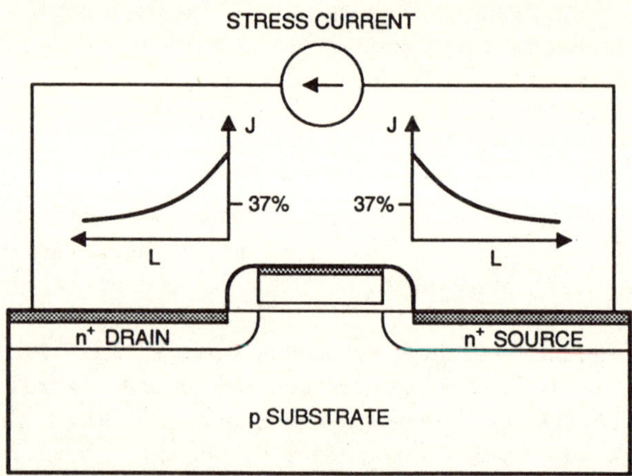

Figure 7.7 Cross-section of an nMOS transistor showing the current density, J, peaking at the silicide edge (after [Rountree88]).

current to flow more in the diffusion, and make for more uniform current densities at the silicide edge. By moving the current into the diffusion earlier, the effective spreading resistance of the contact is increased, thus improving the ballasting effect of the drain. Processing techniques which increase the silicide sheet resistance while reducing the silicide to diffusion contact resistance will force the current into the diffusion earlier and increase the spreading resistance. Additionally, increasing the junction depth will also increase the spreading resistance of the diffusion region thereby increasing the ESD robustness.

The actual effect of the silicide processing will vary for different processes since the doping concentrations of the source and drain diffusions and the anneal times will have an impact on the silicide thickness and the effective resistance [Scott87]. Chen showed that for his process, the ESD performance was best for a 600 Å starting Ti thickness which resulted in $\approx$475 Å of $TiSi_2$ assuming 75% of the Ti is converted into $TiSi_2$. Figure 7.8 shows the dependence of HBM ESD thresholds on the $TiSi_2$ thickness for nMOS transistors. For a Ti thickness of 600 Å the ESD voltage was 4.5 kV. In comparison, for an 800 Å starting Ti thickness the ESD pass voltage was 4 kV and a 1000 Å starting thickness had an ESD pass threshold of 1.5 kV. The cause of the decrease in ESD threshold could be related to the unsilicided silicon junction depth as shown in Figure 7.9. The unsilicided junction depth is calculated by subtracting the amount of silicon consumed during the silicidation process from the original junction depth, x_j, obtained from SUPREM-3 simulations. A minimum silicon junction depth was required to ensure good ESD performance. Thicker silicide also means a lower contact resistance and less ballasting in the drain which would also result in lower ESD robustness. As technologies scale and effective junction depths are reduced, the effects of silicidation on junction depths and contact resistance will be an important issue.

The most effective solution to the silicide problem is to block the silicide from areas close to the gate edge. Ideally, the non-silicided area (silicide block) must be of

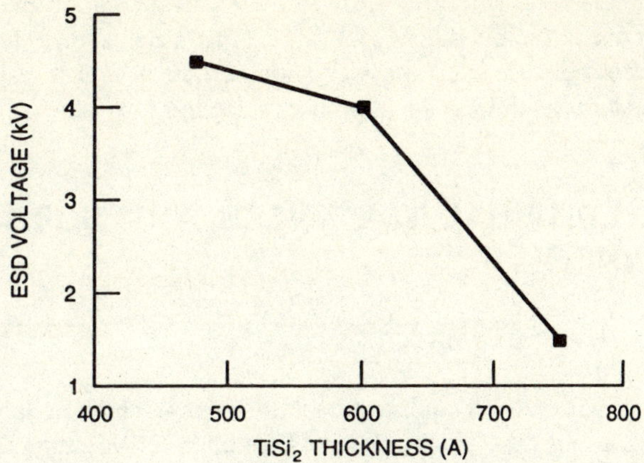

Figure 7.8 Effect of TiSi$_2$ thickness on the ESD pass voltage for nMOS transistors [Chen88].

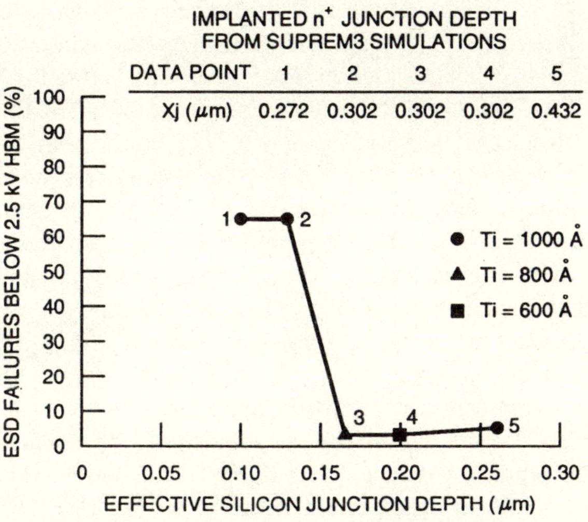

Figure 7.9 Percentage of devices that fail below 2.5 kV HBM as a function of the calculated silicon junction depth (after [Chen88]).

the order of 2 μm to 3 μm to take advantage of the increased drain-contact resistance. Silicide block of ≤ 1 μm has been shown to be ineffective in some instances because it does not provide sufficient ballast resistance to prevent current localization as the device approaches second breakdown [Amerasekera92].

7.5 WELLS, EPITAXIAL THICKNESS AND SUBSTRATE RESISTANCE

7.5.1 Wells

Well concentrations and epitaxial thicknesses have a large impact on protection circuits using dual-diode schemes [Voldman92][Voldman93]. Figure 7.10 shows the ESD performance as a function of the n-well sheet resistance for a dual-diode protection circuit [Voldman92]. As the sheet resistance is decreased from 1100 $\Omega/\square$ to 330 $\Omega/\square$, the ESD performance increases from 2.5 kV to nearly 7 kV. However, higher n-well doping will have an effect on junction capacitance and influence circuit speed. The n-well design must, therefore, be optimized for both ESD performance and transistor/circuit performance.

In nMOS protection circuits, the p-well doping will have an effect on the performance of the lateral npn. Higher p-well doping will make it harder for the lateral npn to turn on, and hence degrade ESD performance. There is as yet not enough information available in the literature to comment on the possible optimum doping levels for the p-well. It would vary depending on the type of source/drain (e.g. LDD, MDD, abrupt) and the junction depth. One important factor caused by the dependence on the p-well doping is that it makes it difficult to directly compare results from different workers. It is necessary to be aware of the possible implications of the p-well doping concentration when comparing published results.

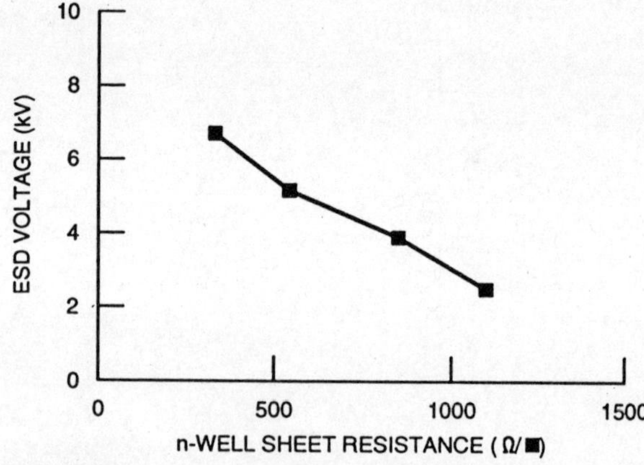

Figure 7.10 ESD performance of a dual-diode protection circuit as a function of n-well sheet resistance [Voldman93].

In SCR protection circuits, increasing the n-well doping will have some effect on the ability to trigger the SCR and will also increase the SCR holding voltage. Hence, the ESD performance of the SCR protection circuit will be affected, particularly with regard to the low level protection capability [Duvvury88].

7.5.2 Epitaxial thickness

The epitaxial thickness can significantly impact the ESD performance. Effects of epitaxial thickness variations have been extensively studied for dual-diode protection circuits [Voldman93] but not much published work is available for other protection device types. The results of HBM ESD thresholds and MM ESD thresholds for dual-diode circuits are shown in Figure 7.11 [Voldman93]. Note that the results will be dependent on the n-well and p-well processing and may not be directly comparable with other processes. The figure shows that as the epitaxial thickness for this process is increased from 1.9 μm to 2.5 μm, the p^+ substrate dopant compensates the n-well doping and changes the n-well sheet resistance. As the n-well resistance drops from 955 $\Omega/\square$ to 550 $\Omega/\square$, the HBM ESD threshold voltage increases from 3.8 kV to 5.6 kV and the MM ESD threshold goes from 700 V to 1.2 kV. Both of these increases are significant amounts. Diodes in the p-well on p-substrates will also be affected by changes in the epitaxial thickness. The problem here is that as the thickness is increased, the diode resistance may also increase, depending on the layout. If adequate substrate contacts are not made, the diode resistance will be dominated by the resistance of the epi layer, which will lower the effectiveness of these protection circuit elements.

The nMOS protection will also see degradation in ESD performance as the epitaxial thickness is decreased. One of the major factors here is the reduced substrate resistance which increases the snapback trigger voltage. If the snapback trigger voltage is too high, junction burn-out or gate-oxide breakdown may occur before the lateral npn is triggered and the protection circuit can operate as desired.

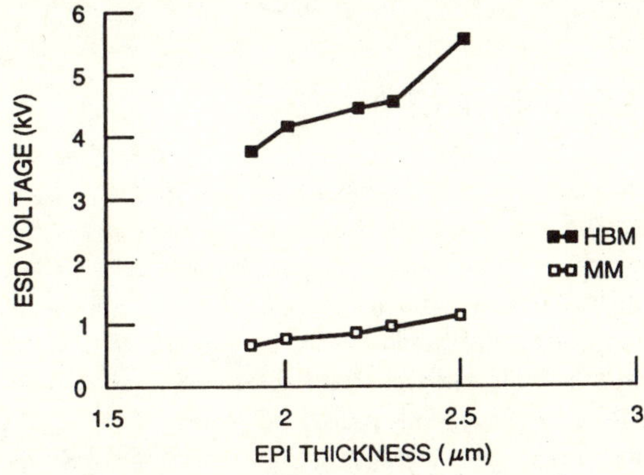

Figure 7.11   ESD performance of a dual-diode protection circuit as a function of epitaxial thickness [Voldman93].

Again, there is very little published data on these effects, and the optimum epitaxial thickness is not defined.

Reducing the epitaxial thickness has the biggest impact on SCR protection circuits. Since one of the primary reasons for using epitaxial layers is to reduce the latchup sensitivity, this is not surprising. The effect comes from the need for a higher trigger current for the SCR, which will mean that the secondary protection circuit elements will be stressed at higher levels and may fail before the SCR triggers [Duvvury88]. In the low voltage SCR, the built-in nMOS transistor will fail if the SCR does not trigger before the current reaches the second breakdown trigger current level for the nMOS. Thinner epi also increases the holding voltage for the SCR (once it triggers). These results indicate that a protection circuit which provided adequate ESD protection in a thicker epi process may result in poor performance when the epitaxial thickness is reduced unless the design is optimized to accommodate the impact of the epitaxial thickness.

7.5.3 Substrate resistance

The substrate resistance is essentially determined by the substrate doping concentration and the epitaxial thickness if used. The major role played by the substrate resistance is in the turn-on of the parasitic bipolar transistor in the nMOS protection circuits [Toyabe78][Schutz82][Hsu82][Laux87]. The substrate resistance being considered here is the spreading resistance between the avalanching drain junction and the highly doped substrate. Higher substrate resistances would be desirable in this case but the performance of the n/substrate diode must also be considered (see Section 7.5.2). Once the bipolar is triggered, the substrate resistance decreases because of conductivity modulation [Laux87]. A higher n/substrate diode resistance will reduce the effectiveness of the nMOS protection for negative ESD stress to substrate. This would be of particular concern in non-epitaxial processes. There is very little actual published information on the effects of substrate resistance on ESD performance and, as in the cases discussed in the sub-sections above, an optimum substrate resistance has not been defined.

7.6 SILICON-ON-INSULATOR (SOI)

SOI technology is at present mostly used for ICs which require high radiation hardness and has wide application in military and space applications. As technologies are scaled to the sub-0.25 μm regime, it is possible that SOI will find application in low-voltage and low-power ICs. The ESD capability of SOI technologies is, therefore, of importance to the IC industry. SOI devices typically have worse ESD performance than devices fabricated in standard bulk CMOS [Verhaege93A] [Chan94]. The main reason for the lower ESD capability of SOI is that the heat dissipation region in the silicon is surrounded by SiO_2. In bulk devices most of the thermal dissipation takes place through the silicon substrate which has a reasonably low thermal resistance. In contrast to the silicon substrates in bulk devices, the buried

SiO_2 has a very low thermal conductivity which results in higher temperatures for the same injected current. However, the floating substrate in SOI devices can provide good triggering of the lateral *npn* transistor which will enable some degree of ESD capability to be achieved in SiO_2 processes.

It has been shown that ESD levels of 10 V/μm can be achieved in SOI nMOS transistors which are comparable to silicided bulk nMOS transistors [Verhaege93B]. The ESD capability is a function of the channel length, and decreases with increasing channel length as has also been observed in bulk devices. One of the main features in SOI devices which enable high ESD levels is the *double snapback* phenomenon described by Verhaege *et al* [Verhaege93A]. Double snapback occurs because of the presence of two parasitic bipolar paths in SOI nMOS devices. The first snapback is associated with the top interface while the second snapback takes place when the potential in the depleted region deeper in the film forms a second bipolar path.

7.7 PACKAGING

The main effect of the package is on the total capacitance of the IC when it is inserted into a test socket or circuit board [Roozendaal90][Verhaege93C]. There has also been some concern regarding the effects of ceramic compared to plastic packaging on the ESD performance of a chip. Unpublished work by the authors has indicated that, provided the package size remains constant (e.g. 40 pin dual-in-line, DIL), the ESD performance of a given chip does not change between these package types. However, if the comparison is made between a DIL and a thin quad flat package (TQFP) then it is possible that differences are observed. The same is true of the difference between a 40 pin DIL and a 16 pin DIL both in plastic [Roozendaal90]. The effect of the package capacitance is to increase the rise time of the applied ESD pulse, which can have an impact on the ESD performance of the device. Whether the performance improves or degrades depends on the protection circuit design used [Verhaege93C].

One of the significant effects of the package and testboard capacitance is that it is charged up during the ESD event and subsequently discharges when the ESD stress is over [Roozendaal90][Verhaege93C]. This effectively generates a second current stress in the ESD protection circuit. The discharge circuit of the second current pulse has very little series resistance, and the series inductance is solely determined by the package interconnects and the testboard tracks. This can lead to peak currents two to three times higher than the original stress current depending on the magnitude of the capacitances and inductances involved. In general, therefore, devices packaged in high capacitance packages may show a higher sensitivity to ESD damage than those in a lower capacitance package. However, the sensitivity will depend on the type of protection used, i.e. a fast switching protection will be less sensitive than a slow switching protection. In the case of a fast switching circuit it is possible that the smaller capacitance package will result in a higher ESD sensitivity.

With the increased usage of the Charged Device Model type test method (see Chapter 2), the effect of the package size and whether it is ceramic or plastic will become more important to the ESD performance [Verhaege94][Gieser94].

REFERENCES

[Amerasekera86]   A. Amerasekera, D. Campbell, 'ESD Pulse and Continuous Voltage Breakdown in MOS Capacitor Structures', in *Proc. 8th EOS/ESD Symposium*, p. 208–213, 1986.

[Amerasekera90]   A. Amerasekera, L. J. van Roozendaal, J. Abderhalden, J. J. P. Bruines, L. Sevat, 'An Analysis of Low Voltage ESD Damage in Advanced CMOS Processes', in *Proc. 12th EOS/ESD Symposium*, p. 143–150, 1990.

[Amerasekera91]   A. Amerasekera, L. van Roozendaal, J. Bruines, F. Kuper, 'Characterization and Modeling of Second Breakdown in NMOST's for the Extraction of ESD-Related Process and Design Parameters', *IEEE Trans. Elec. Dev.*, ED-38, p. 2161–2168, 1991.

[Amerasekera92]   A. Amerasekera, W. van den Abeelen, L. J. van Roozendaal, M. Hannemann, P. J. Schofield, 'ESD Failure Modes: Characteristics, Mechanisms and Process Influences', *IEEE Trans. Elec. Dev.*, ED-39, p. 2, 1992.

[Amerasekera94A]   A. Amerasekera, R. Chapman, 'Technology Design for High Current and ESD Robustness in a Deep Submicron Process', *Elec. Dev. Lett.*, 15, p. 383–385, 1994.

[Amerasekera94B]   A. Amerasekera, C. Duvvury, 'The Impact of Technology Scaling on ESD Robustness and Protection Circuit Design', in *Proc. 16th EOS/ESD Symposium*, 1994.

[Amerasekera94C]   A. Amerasekera, J. A. Seitchik, 'Electrothermal Behavior of nMOS Transistors under High Current Snapback (ESD/EOS) Conditions', in *Tech. Dig. IEDM*, p. 446–449, 1994.

[Bridgewood85]   M. Bridgewood, R. Kelly, 'Modeling the Effects of Narrow Impulsive Overstress on Capacitive Test Structures', in *Proc. 7th EOS/ESD Symposium*, p. 84–91, 1985.

[Chaine92]   M. Chaine, S. Desai, C. Dunn, D. Dolby, W. Holland, T. Pekny, 'ESD Improvement Using Low Concentrations of Arsenic Implantation in CMOS Output Buffers', in *Proc. 14th EOS/ESD Symposium*, p. 136–142, 1992.

[Chan94]   M. Chan, S. S. Yuen, Z.-J. Ma, K. Y. Hui, P. K. Ko, C. Hu, 'Comparison of ESD Protection Capability of SOI and Bulk CMOS Output Buffers', in *Proc. 32nd IRPS*, p. 292–298, 1994.

[Chang88]   P.-H. Chang, R. Hawkins, T. Bonifield, L. Melton, 'Aluminum Spiking at Contact Windows in Al/Ti–W/Si', *App. Phys. Lett.*, 52, p. 272–274, 1988.

[Chen86]   K. L. Chen, G. Giles, D. Scott, 'Electrostatic Discharge Protection for 1-μm CMOS Devices and Circuits', in *Tech. Dig. IEDM*, p. 484–487, 1986.

[Chen88]   K.-L. Chen, 'Effect of Interconnect Process and Snapback Voltage on the ESD Failure Threshold of NMOS Transistors', in *Proc. 10th EOS/ESD Symposium*, p. 212–219, 1988.

[Dabral93]   S. Dabral, R. Aslett, 'Designing On-Chip Power Supply Coupling Diodes for ESD Protection and Noise Immunity', in *Proc. 15th EOS/ESD Symposium*, p. 239–250, 1993.

[Daniel90]   S. Daniel, G. Krieger, 'Process and Design Optimization for Advanced CMOS I/O ESD Protection Devices', in *Proc. 12th EOS/ESD Symposium*, p. 206–213, 1990.

[DeChiaro81]   L. F. De Chiaro, 'Electrothermomigration in nMOS LSI Devices', in *Proc. 19th IRPS*, p. 212–217, 1981.

[Duvvury85]   C. Duvvury, R. Rountree, D. Baglee, A. Hyslop, L. White, 'ESD Design Considerations for ULSI', in *Proc. 7th EOS/ESD Symposium*, p. 45–48, 1985.

[Duvvury86]   C. Duvvury, R. McPhee, D. Baglee, R. Rountree, 'ESD Protection Reliability in 1-μm CMOS Circuit Performance', in *Proc. 24th IRPS*, p. 199–208, 1986.

[Duvvury88]   C. Duvvury, R. Rountree, O. Adams, 'Internal Chip ESD Phenomena Beyond the Protection Circuit', in *Proc. 26th IRPS*, p. 19–25, 1988.

[Duvvury89]   C. Duvvury, R. Rountree, H. Stiegler, T. Polgreen, D. Corum, 'ESD Phenomenon in Graded Junction Devices', in *Proc. 27th IRPS*, p. 71–76, 1989.

[Fong87]   Y.-P. Fong, C. Hu, 'The Effects of High Electric Field Transients on Thin Gate Oxide MOSFETs', in *Proc. 9th EOS/ESD Symposium*, p. 252–257, 1987.

[Fukuda86]   Y. Fukuda, S. Ishiguro, M. Takahara, 'ESD Protection Network Evaluation by HBM and CPM (Charged Package Method)', in *Proc. 8th EOS/ESD Symposium*, p. 193–199, 1986.

[Gieser94]   H. Gieser, P. Egger, 'Influence of Tester Parasitics on Charged Device Model Failure Thresholds', in *Proc. 16th EOS/ESD Symposium*, p. 69–84, 1994.

[Hsu82]   F. C. Hsu, P. K. Ko, S. Tam, C. Hu, R. Muller, 'An Analytical Breakdown Model for Short-Channel MOSFETs', *IEEE Trans. Elec. Dev.*, ED-29, p. 1735–1740, 1982.

[Ishizuka94]   H. Ishizuka, K. Okuyama, K. Kubota, 'Photon Emission Study of ESD Protection Devices under Second Breakdown Conditions', in *Proc. 32nd IRPS*, p. 286–291, 1994.

[Lau82]   C. K. Lau, Y. C. See, D. Scott, J. Bridges, S. Perna, R. D. Davies, 'Titanium Disilicide Self-Aligned Source/Drain and Gate CMOS Technology', in *Tech. Dig. IEDM*, p. 714–717, 1982.

[Laux87]   S. E. Laux, F. H. Gaensslen, 'A Study of Channel Avalanche Breakdown in Scaled n-MOSFETs', *IEEE Trans. Elec. Dev.*, ED-34, p. 1066–1073, 1987.

[Maloney86]   T. Maloney, 'Contact Injection: a Major Cause of ESD Failure in Integrated Circuits', in *Proc. 8th EOS/ESD Symposium*, p. 166–172, 1986.

[McPhee86]   R. McPhee, C. Duvvury, R. N. Rountree, 'Thick Oxide Device ESD Performance under Process Variations', in *Proc. 8th EOS/ESD Symposium*, p. 173–181, 1986.

[Ohtani90]   S. Ohtani, M. Yoshida, 'Model of Leakage Current in LDD Output MOSFET Due to Low-Level ESD Stress', in *Proc. 12th EOS/ESD Symposium*, p. 177–181, 1990.

[Pierce85]   D. G. Pierce, 'Electro-Thermomigration as an Electrical Overstress Failure Mechanism', in *Proc. 7th EOS/ESD Symposium*, p. 67–76, 1985.

[Rafferty93]   C. S. Rafferty, H.-H. Vuong, S. A. Eshraghi, M. Giles, M. Pinto, S. Hillenius, 'Explanation of Reverse Short-Channel Effect by Defect Gradients', in *Tech. Dig. IEDM*, p. 311–314, 1993.

[Roozendaal90]   L. J. van Roozendaal, A. Amerasekera, P. Bos, W. Baelde, F. Bontekoe, P. Kersten, E. Korma, P. Rommers, P. Krys, U. Weber, P. Ashby, 'Standard ESD Testing', in *Proc. 12th EOS/ESD Symposium*, p. 119–130, 1990.

[Rountree88]   R. Rountree, 'ESD Protection for Submicron CMOS Circuits Issues and Solutions', in *Tech. Dig. IEDM*, p. 580–583, 1988.

[Scott86]   D. Scott, J. Hall, G. Giles, 'A Lumped Element Model for Simulation of ESD Failures in Silicided Devices', in *Proc. 8th EOS/ESD Symposium*, p. 41–47, 1986.

[Scott87]   D. Scott, R. Chapman, C.-C. Wei, S. S. Mahant-Shetti, R. Haken, T. Holloway, 'Titanium Disilicide Contact Resistivity and Its Impact on 1-μm CMOS Circuit Performance', *IEEE Trans. Elec. Dev.*, ED-34, p. 562–574, 1987.

[Schutz82]   A. Schutz, S. Selberherr, H. Potzl, 'Analysis of Breakdown in MOSFETs', *IEEE Comp. Aid. Des.*, CAD-1, p. 77–84, 1982.

[Shabde84]   S. N. Shabde, G. Simmons, A. Baluni, D. Back, 'Snapback Induced Gate Dielectric Breakdown in Graded Junction MOS Structures', *Proc. 22nd IRPS*, p. 165–168, 1984.

[Strauss87]   M. Strauss, D. Lin, T. Welsher, 'Variations in Failure Modes and Cumulative Effects Produced by Commercial Human Body Model ESD Simulators', in *Proc. 9th EOS/ESD Symposium*, p. 59–63, 1987.

[Toyabe78]   T. Toyabe, K. Yamaguchi, S. Asai, M. Mock, 'A Two-Dimensional Avalanche Breakdown Model of Submicron MOSFETs', *IEEE Trans. Elec. Dev.*, ED-25, p. 825–833, 1978.

[Verhaege93A]   K. Verhaege, G. Groeseneken, J.-P. Collinge, H. E. Maes, 'Double Snapback in SOI nMOSFETs and its Application for SOI ESD Protection', *Elec. Dev. Lett.*, EDL-14, p. 326–328, 1993.

[Verhaege93B]   K. Verhaege, G. Groeseneken, J. P. Collinge, H. E. Maes, 'The ESD Protection Capability of SOI Snapback nMOS FETs: Mechanisms and Failure Modes', in *Proc. 15th EOS/ESD Symposium*, p. 215–219, 1993.

[Verhaege93C]   K. Verhaege, P. Roussel, G. Groeseneken, H. Maes, H. Gieser, C. Russ, P. Egger, X. Guggenmos, F. Kuper, 'Analysis of HBM ESD Testers and Specifications Using 4th Order Lumped Element Model', in *Proc. 15th EOS/ESD Symposium*, p. 129–138, 1993.

[Verhaege94]   K. Verhaege, G. Groeseneken, H. Maes, P. Egger, H. Gieser, 'Influence of Tester, Test Method and Device Type on CDM ESD Testing', in *Proc. 16th EOS/ESD Symposium*, p. 49–62, 1994.

[Voldman92]   S. Voldman, V. Gross, M. J. Hargrove, J. M. Never, J. A. Slinkman, M. P. O'Boyle, T. S. Scott, J. J. Delecki, 'Shallow Trench Isolation Double-Diode Electrostatic Discharge Circuit and Interaction with DRAM Output Circuitry', in *Proc. 14th EOS/ESD Symposium*, p. 277–288, 1992.

[Voldman93]   S. Voldman, V. Gross, 'Scaling Optimization and Design Considerations of Electrostatic Discharge Protection Circuits in CMOS Technology', in *Proc. 15th EOS/ESD Symposium*, p. 251–260, 1993.

[Voldman94]   S. Voldman *et al*, 'ESD Protection in a Multi-Rail Disconnected Power Grid and Mixed Voltage Interface Environments in 0.5 and 0.25 μm Channel Length CMOS Technologies', in *Proc. 16th EOS/ESD Symposium*, 1994.

[Wei92]   Y. Wei, Y. Loh, C. Wang, 'MOSFET Drain Engineering for ESD Performance', in *Proc. 14th EOS/ESD Symposium*, p. 143–148, 1992.

[Wilson87]   D. Wilson, H. Domingos, M. M. S. Hassan, 'Electrical Overstress in nMOS Silicided Devices', in *Proc. 9th EOS/ESD Symposium*, p. 265–273, 1987.

8

CONCLUSIONS

8.1 LONG-TERM RELEVANCE OF ESD IN ICS

ESD damage is directly responsible for approximately 10% of the total failure returns [Green88][Wagner93] and the reported number may be higher but for the difficulty in distinguishing between some EOS and ESD failures. This makes it an important failure mechanism throughout the semiconductor industry. As technology feature sizes move into the deep submicron regime, concerns regarding ESD are increasing. Shallower junctions, very thin gate oxides and small channel lengths associated with technology scaling will effect ESD performance, although the indications are that the impact may be positive rather than negative [Lin93][Amerasekera94]. However, together with the smaller feature sizes, future generations of ICs will also consist of very high density circuits with pad counts in excess of 500. This will cause a severe crunch on the available area for ESD protection circuitry requiring that the protection circuits become more efficient.

The impact of ESD damage due to handling and testing can have a negative influence on product yield [Wagner93]. Large ICs manufactured in advanced processes may only have 30 to 40 chips per 6 inch wafer. Each of these dies can eventually sell for anything between $300 and $3000 depending on the application and the availability of alternative products. Any product loss due to ESD damage has a direct impact on profitability and even fall-outs of the order of 1% are not acceptable. There is strong motivation, therefore, to ensure that the present and future ICs have reasonable ESD levels to avoid damage during handling and testing. Another issue which gives increasing importance to ESD is the move towards replaceable ICs in electronic systems. Instead of replacing the whole circuit board, as used to be standard practice, users are encouraged to purchase upgrades to their microprocessors and memory cards and do the installation themselves. Since the installation does not neccessarily take place in an ESD-safe environment, the ICs need to be ESD robust.

The demand for ESD robustness has led to more consideration for ESD robustness during technology development and circuit design. At the same time ESD test methods are being better defined and correlated to real ESD events, thus improving the confidence levels in the ability of protection circuits to function as required.

8.2 STATE-OF-THE-ART FOR ESD PROTECTION

Table 8.1 and Table 8.2 list the present ESD protection circuit elements for MOS processes and bipolar/BiCMOS processes respectively, together with the *trigger voltage* and the *clamping voltage*. The most widely used ESD protection circuits are based on the high current *npn* transistor action in both bipolar/BiCMOS and CMOS technologies. The use of SCR structures is increasing because of their lower power dissipation after they trigger, but the high trigger voltages have limited their application. Reducing the trigger voltages for both SCR as well as *npn* devices has been a major thrust in the present development of ESD protection circuits with reasonable success.

The high chip capacitance in large ICs has increased the effectiveness of the diode protection circuit which was popular in the early technologies. A diode between the pad and the positive power supply (V_{cc}) is usually supplemented by either an *npn* device or an SCR to the ground (V_{ss}) or negative supply. Diode circuits also require good protection between V_{cc} and ground which can be an *npn* type circuit or an SCR depending on the application.

Table 8.1 ESD protection circuit elements for MOS processes.

ELEMENT	TRIGGER VOLTAGE	CLAMPING VOLTAGE	COMMENTS
DIODE	15 TO 20V	15 TO 20V	HIGH "ON" RESISTANCE
GROUNDED GATE NMOST	12 TO 15V	6 TO 8V	LOW POWER DISSIPATION CAPABILITY
THICK OXIDE MOS	15 TO 20V	8 TO 12V	HIGH TRIGGER VOLTAGE
SILICON-CONTROLLED RECTIFIERS (SCRs)	12 TO 25V	1 TO 5V	HIGH TRIGGER VOLTAGE; LOW "ON" RESISTANCE

Table 8.2 ESD protection circuit elements for bipolar/BiCMOS processes.

ELEMENT	TRIGGER VOLTAGE	CLAMPING VOLTAGE	COMMENTS
npn	18 TO 22V	8 TO 10V	HIGH TRIGGER
BIPOLAR SCR	20 TO 25V	1 TO 5V	HIGH TRIGGER

IN ADDITION, DIODES AND MOS ELEMENTS CAN BE USED IN BiCMOS PROCESSES.

Large ICs have a large number of pin combinations and consideration must be given to each of these when placing ESD protection circuits. In addition, in many cases multiple V_{cc} and V_{ss} buses may be used which lead to added complications in terms of protection circuitry. In general, ESD protection circuits are placed between all the important combinations. This requirement makes the use of diode elements attractive because of their small areas.

The effect of technology on ESD protection circuit performance has been a major hindrance to achieving consistent ESD behavior. Advanced process development has begun to include ESD considerations in the technology roadmap. In some cases process features have been implemented just to ensure consistent ESD performance.

It is our opinion that the state-of-the-art still leaves much to be done towards achieving good ESD protection circuit design methodologies for consistent ESD performance. Some of the current limitations and the future issues involved will be discussed in the rest of this chapter.

8.3 CURRENT LIMITATIONS

The present approach to ESD protection circuit design and implementation is iterative. Circuits are designed and evaluated depending on the available area and the pin specifications. In many cases the ESD capability of the process has been previously characterized on test structures to enable basic design guidelines for the protection circuits to be generated. However, in many cases the iterative approach involves too much time and can result in a delay in the release of a product to the market. Since a delay could be costly, the outcome is that ESD protection circuit performance may be sacrificed.

There is a need, therefore, to reduce the number of cycles required for the development of good ESD protection circuits. An important contribution in this direction would be the ability to use accurate simulation tools to evaluate the circuit and the technology before committing to silicon. Present generations of ICs are almost entirely designed using simulations, and the inability to include the ESD circuit into the simulation loop makes it more difficult to include ESD performance into the circuit design. The same is true for technology design. The lack of accurate simulation tools is one of the major reasons why ESD is considered to be almost a 'black art' in the semiconductor industry.

Chapter 6 presented a summary of the state-of-the-art of modeling ESD events in semiconductors. It shows that there has been progress made towards the development of suitable ESD simulators. The two main obstacles to progress in the areas of circuit and technology simulation are the limited understanding of

(a) the mechanisms governing the technology dependence of ESD sensitivity; and
(b) the circuit interactions during an ESD event.

A further limitation is the inability to automatically check a full-circuit design with regard to whether it follows all the ESD design rules. Many ESD problems in ICs are due to errors in the layout which lead to a low ESD damage threshold although the protection circuit is well designed and can pass the required ESD levels itself. An ESD design rule checker would have to be able to search for and identify transistors which would be directly in the path of an ESD current stress and do not have adequate ESD protection. It would also need to ensure that unrelated diffusions would not trigger parasitic *npn* or SCR devices during an ESD event.

The present range of protection circuits all depend on the triggering of parasitic elements in order to provide suitable levels of ESD protection. Although these parasitic elements have successfully enabled very high ESD levels to be reached, it is difficult to ensure consistent triggering of these devices under all ESD conditions. At present this is not a major limitation, but with the advent of the CDM test method it is foreseeable that circuits which consistently work under a range of stresses would be needed. Such circuits may very well be active circuits whose operation is not based on the triggering of parasitic devices.

On a different note, almost all ICs, with the possible exception of a few high voltage devices used in automotive applications, use on-chip ESD protection techniques. As ICs become large and technologies become smaller there may be an advantage in moving towards off-chip ESD protection circuits [Cronin93][Unger94][Lin94]. Such protection may use techniques such as crowbars which short all pins together when not inserted into a socket, thus preventing a pin-to-pin type ESD stress from taking place. One of the concerns in off-chip protection techniques is that the dynamic impedance of the 'short' during a fast ESD event must be low [Lin94]. It is possible that even though the DC impedance is close to zero, the dynamic impedance during a CDM event can be large because of the high dI/dt. The governing parameter is the inductance of the wire or metalization used for the current shunt and this should be as small as possible.

8.4 FUTURE ISSUES

The current limitations listed in the previous section would form the basis of the main issues to be addressed as technologies move towards sub-0.25 μm feature sizes. It is essential to improve our understanding of the underlying mechanisms of the behavior of circuit elements during an ESD event. We would then be able to develop consistent predictive methods which can be used to evaluate the impact of technology variations on ESD performance. In any case, ESD requirements should be part of the process development roadmap to ensure that good ESD performance is achieved in future generations of ICs.

A parallel action would be the development of circuit simulators which incorporate the capability to extract the sensitivity to both HBM and CDM type ESD stress events. At present it is expected that such a simulator would require a large amount of memory and computing power because of the large number of parasitics involved. For example, in a large IC each ESD event could select a different path between each of the power buses. These paths need to be accurately simulated to determine the chip layout factors influencing the ESD capability of these preferred paths or to channel the stress current through a dedicated ESD protection circuit.

The need for a good ESD design rule checker is clear. Such a checker needs to be able to consistently identify ESD sensitive paths and violations of ESD design guidelines to be effective. Once again, it is not certain whether the computational power required to accomplish such a task is available in the near future.

Attempts to develop new and improved ESD protection circuits are unlimited. Most of the present interest has been on better ways of triggering these protection devices at lower voltages. There is a move towards using the path between the V_{cc} and V_{ss} buses as the primary protection in large ICs and this approach may become more common in the next generation of ICs [Merrill93][Voldman94A][Voldman94B][Tandan94][Croft94][Dabral94]. As ICs become larger and more complex and on-chip area becomes more valuable, off-chip protection techniques may gain popularity. However, this would not eliminate the need to design robust circuits or technologies, but it will mean that large area on-chip protection circuits will not be used.

One area of protection circuit design which will gain importance in the coming generations of ICs is the protection of high speed telecom chips. ICs with bit rates exceeding 2 GBits/s cannot tolerate much capacitive or resistive loading due to the ESD protection circuits. Some of the trade-offs involved between ESD protection and normal operation have been discussed in Chapter 4. However, there is a need for a protection circuit design methodology which allows these high speed circuits to be made self-protecting if possible. The alternative is to use off-chip protection techniques.

A second area of importance in protection circuit design for the present and future generations of ICs is the use of multi-voltage supplies, i.e. 3.3 V/5 V, or 2.5 V/3.3 V, in these chips (see e.g. [Voldman94A][Voldman94B]). The requirement that a circuit using elements designed for 3.3 V operation is able to tolerate 5 V means that in some cases it is not possible to use protection circuits where nMOS devices are the primary protection elements. New techniques are being developed for these ICs [Voldman94B]. However, there are many new challenges to be faced in the design of efficient and optimized ESD protection circuits for multi-voltage applications.

Finally, the complexities and intricacies involved in ESD in silicon integrated circuits have generated interest in many academic and industrial research workers. A systematic approach towards understanding the fundamentals of the issues involved will eventually lead to solutions to some of the issues raised in this chapter. In the last 15 years large strides have been made towards understanding and solving some of the most pressing ESD problems. There are still a number of issues which, while making ESD in ICs an interesting and rewarding (if sometimes frustrating!) field for research, need to be resolved if good, consistent ESD protection is to be designed for future generations of ICs.

REFERENCES

[Amerasekera94]   A. Amerasekera, C. Duvvury, 'The Impact of Technology Scaling on ESD Robustness and Protection Circuit Design', in *Proc. 16th EOS/ESD Symposium*, p. 237–245, 1994.

[Croft94]   G. Croft, 'ESD Protection Using a Variable Voltage Supply Clamp', in *Proc. 16th EOS/ESD Symposium*, p. 135–140, 1994.

[Cronin93]   D. Cronin, 'CRO-BAR: a New Technique for ESD Protection', *EMC Test and Design*, January 1993.

[Dabral94]   S. Dabral, R. Aslett, T. Maloney, 'Core Clamps for Low Voltage Technologies', in *Proc. 16th EOS/ESD Symposium*, p. 141–148, 1994.

[Green88]   T. Green, 'A Review of EOS/ESD Field Failures in Military Equipment', in *Proc. 10th EOS/ESD Symposium*, p. 7–14, 1988.

[Lin93]   D. L. Lin, 'ESD Sensitivity and VLSI Technology Trends; Thermal Breakdown and Dielectric Breakdown', in *Proc. 15th EOS/ESD Symposium*, p. 73–82, 1993.

[Lin94]   D. Lin, M.-C. Jon, 'Off-Chip Protection: Shunting of ESD Current by Metal Traces on Circuit Packs', in *Proc. 16th EOS/ESD Symposium*, p. 279–285, 1994.

[Merrill93]   R. Merrill, E. Issaq, 'ESD Design Methodology', in *Proc. 15th EOS/ESD Symposium*, p. 233–237, 1993.

[Tandan94]   N. Tandan, G. Conner, 'ESD Trigger Circuit', in *Proc. 16th EOS/ESD Symposium*, p. 120–124, 1994.

[Unger94]   B. Unger, private communication, 1994.

[Voldman94A]   S. Voldman, 'ESD Protection in a Multi-Rail Disconnected Power Grid and Mixed Voltage Interface Environments in 0.5 and 0.25 μm Channel Length CMOS Technologies', in *Proc. 16th EOS/ESD Symposium*, p. 125–134, 1994.

[Voldman94B]   S. Voldman, G. Gerosa, 'Mixed-Voltage-Interface ESD Protection Circuits for Advanced Microprocessors in Shallow Trench Isolation and LOCOS CMOS Technology', in *Tech. Dig. IEDM*, 1994.

[Wagner93]   R. G. Wagner, J. Soden, C. F. Hawkins, 'Extent and Cost of EOS/ESD Damage in an IC Manufacturing Process', in *Proc. 15th EOS/ESD Symposium*, p. 49–55, 1993.

AUTHOR INDEX